# Advances in Soil Science

# Advances in Soil Science

B.A. Stewart, Editor

# Advances in Soil Science

## Volume 6

Edited by B.A. Stewart

With Contributions by
R.R. Allmaras, S.C. Gupta, J. Kubota, A.D. Scott,
S.J. Smith, D.L. Sparks, M.E. Sumner, D. Van Campen, J.L. Walworth,
and R.M. Welch

With 65 Illustrations

Springer-Verlag
New York Berlin Heidelberg
London Paris Tokyo

B.A. Stewart
USDA Conservation & Production Research Laboratory
Bushland, Texas 79012
USA

ISSN 0176-9340



Typeset by David Seham Associates, Metuchen, New Jersey.

9 8 7 6 5 4 3 2 1

ISBN-13: 978-1-4612-9112-1 e-ISBN-13: 978-1-4612-4682-4
DOI: 10.1007/978-1-4612-4682-4

# Preface

The world needs for food and fiber continue to increase. Population growth in the developing countries peaked at 2.4 percent a year in 1965 and has fallen to about 2.1 percent. However, in many developing countries almost half the people are under 15 years of age, poised to enter their productive and reproductive years. The challenges to produce enough food for this growing population will remain great. Even more challenging is growing the food in the areas of greatest need. Presently the world has great surpluses of food and fiber in some areas while there are devastating deficiencies in other areas. Economic conditions and the lack of suitable infrastructure for distribution all too often limit the alleviation of hunger even when there are adequate supplies, sometimes even within the country itself.

World hunger can only be solved in the long run by increasing crop production in the areas where the population is growing most rapidly. This will require increased efforts of both the developed and developing countries. Much of the technology that is so successful for crop production in the developed countries cannot be utilized directly in the developing countries. Many of the principles, however, can and must be adapted to the conditions, both physical and economic, of the developing countries.

This series, *Advances in Soil Science,* was established to provide a forum for leading scientists to analyze and summarize the available scientific information on a subject, assessing its importance and identifying additional research needs. But most importantly, the contributors were asked to develop and identify principles that have practical applications to both developing and developed agricultures. It is not the purpose of the series to report new research findings because there are many excellent scientific journals for that need. Communications in scientific journals, however, are generally restricted to short and technical presentations. Therefore, *Advances in Soil Science* fills a gap between the scientific journals and the comprehensive reference books in which scientists can delve in depth on a particular subject relating to soil science.

The ultimate aim of the series is to stimulate action: action to determine where there are arable soils, action to develop technology for more efficient crop production on these soils, action to reduce the risk of degrading these soil resources, and action to determine on which soils our research efforts should be concentrated. Without such action, the task of producing adequate food in the future may simply be too great.

There are many audiences to reach. While intended primarily for scientists and students of soil science, this series also provides technical information for anyone interested in our natural resources and man's influence on these resources. The reviews are written by leading scientists from many countries, and will therefore provide the reader with information from a wide array of conditions. Such information will be particularly useful to professionals working in areas with developing agricultures because the reviews summarize and assess the significance of the technical literature.

The fact that this is the sixth volume of the series is evidence that there was indeed a need, as perceived, for a medium to publish reviews relating to soil science. I wish to acknowledge the authors for their excellent contributions and cooperation. I should also like to thank the members of the Editorial Board for their assistance in selecting such competent and pleasant authors with whom I had the pleasure of working. Lastly, and most importantly, I want to thank the readers for their acceptance and use of *Advances in Soil Science*.

B.A. Stewart

# Contents

**Preface** .......... v

**Potassium Dynamics in Soils** .......... 1
*Donald L. Sparks*
I. Introduction .......... 2
II. General Characteristics of Soil K .......... 4
III. Forms of Soil Potassium .......... 5
IV. Inorganic and Organic Colloids Important in Potassium Dynamics 8
V. Techniques for Measuring Potassium Dynamics .......... 16
VI. Equations and Models to Describe Potassium Dynamics in Soils and Clay Minerals .......... 19
VII. Effect of Temperature on K Dynamics .......... 27
VIII. Dynamics Between Solution and Exchangeable Phases of Potassium .......... 28
IX. Kinetics of Nonexchangeable and Mineral Potassium Release .......... 32
X. Use of Kinetics for Determining Thermodynamics of Potassium Exchange in Soils .......... 44
XI. Summary and Future Research Needs .......... 52
References .......... 52

**Models to Assess the Susceptibility of Soils to Excessive Compaction** .......... 65
*S.C. Gupta and R.R. Allmaras*
I. Introduction .......... 65
II. Definitions .......... 66
III. Modeling Laboratory Soil Compaction .......... 68
IV. Ease/Susceptibility of Soils to Compaction .......... 77
V. Excessive Compaction .......... 78
VI. Water Contents and Applied Stresses Not Conducive to Excessive Compaction .......... 83
VII. Acceptable Loads Versus Drainage .......... 84

VIII. Modeling Field Soil Compaction .... 86
IX. Application of Field Soil Compaction Models .... 94
X. Areas of Future Research .... 97
References .... 98

**Sources, Amounts, and Forms of Alkali Elements in Soils** .... 101
*A.D. Scott and S.J. Smith*
I. Introduction .... 101
II. Differentiating Characteristics .... 102
III. Geochemistry .... 107
IV. Mineralogy .... 113
V. Soil Content .... 117
VI. Soil Forms .... 124
References .... 138

**The Diagnosis and Recommendation Integrated System (DRIS)** .... 149
*J.L. Walworth and M.E. Sumner*
I. Introduction .... 149
II. Nutrient Concentration and Aging .... 150
III. DRIS Norms .... 154
IV. Making a Diagnosis: Use of a DRIS Chart .... 158
V. Calculating DRIS Indices .... 160
VI. Nutrient Index Interpretation .... 163
VII. Testing DRIS Norms .... 166
VIII. Comparisons of DRIS and Other Diagnostic Systems .... 170
IX. Effect of Leaf Age and Position on DRIS Indices .... 175
X. Universality of Foliar Norms .... 179
XI. Expansion of DRIS Beyond Nutrient Ratios .... 182
XII. Conclusions .... 184
XIII. References .... 185

**Soil-Related Nutritional Problem Areas for Grazing Animals** .... 189
*J. Kubota, R.M. Welch, and D. Van Campen*
I. Introduction .... 189
II. General Considerations of Trace Element Problems in Plants and Animals .... 190
III. Geographical Distribution of Nutritional Problem Areas for Grazing Animals .... 194
IV. General Discussion .... 205
V. Future Outlook .... 211
References .... 212

**Index** .... 217

# Contributors


R.R. ALLMARAS, Soil and Water Management Unit, USDA-ARS, St. Paul, Minnesota 55108, USA

S.C. GUPTA, Department of Soil Science, University of Minnesota, St. Paul, Minnesota 55108, USA

J. KUBOTA, U.S. Plant, Soil and Nutrition Laboratory, Ithaca, New York, and the Department of Agronomy, Cornell University, Ithaca, New York 14853, USA

A.D. SCOTT, Department of Agronomy, Iowa State University, Ames, Iowa 50011, USA

S.J. SMITH, Agricultural Research Service, U.S. Department of Agriculture, Durant, Oklahoma 74702, USA

D.L. SPARKS, Department of Plant Science, University of Delaware, Newark, Delaware 19717, USA

M.E. SUMNER, Department of Agronomy, College of Agriculture, The University of Georgia, Athens, Georgia 30602, USA

D. VAN CAMPEN, U.S. Plant, Soil and Nutrition Laboratory, Ithaca, New York, and the Department of Agronomy, Cornell University, Ithaca, New York 14853, USA

J.L. WALWORTH, Department of Soils and Crops, Rutgers University, New Brunswick, New Jersey 08903, USA

R.M. WELCH, U.S. Plant, Soil and Nutrition Laboratory, Ithaca, New York, and the Department of Agronomy, Cornell University, Ithaca, New York 14853, USA

# Potassium Dynamics in Soils

Donald L. Sparks*


I. Introduction .......... 2
II. General Characteristics of Soil K .......... 4
III. Forms of Soil Potassium .......... 5
IV. Inorganic and Organic Colloids Important in Potassium Dynamics .......... 8
  A. Micas .......... 8
  B. Feldspars .......... 10
  C. Vermiculite .......... 12
  D. Allophane .......... 13
  E. Zeolites .......... 13
  F. Smectites .......... 14
  G. Chlorites and Hydroxy-Interlayered Expansible Phyllosilicates .......... 14
  H. Kaolinite and Halloysite .......... 15
  I. Potassium-Taranakite .......... 15
  J. Potassium-Alunite .......... 15
  K. Organic Matter .......... 16
V. Techniques for Measuring Potassium Dynamics .......... 16
  A. Batch Techniques .......... 16
  B. Flow or Miscible Displacement Techniques .......... 17
  C. Critical Evaluation of Kinetic Techniques .......... 17
VI. Equations and Models to Describe Potassium Dynamics in Soils and Clay Minerals .......... 19
  A. Basic Equations .......... 19
  B. Kinetic Models for Potassium .......... 24
VII. Effect of Temperature on K Dynamics .......... 27
VIII. Dynamics Between Solution and Exchangeable Phases of Potassium .......... 28
  A. Historical Perspective .......... 28
  B. Rate of Exchange on Clay Minerals .......... 29


---

*Associate Professor of Soil Chemistry, Department of Plant Science, University of Delaware, Newark, Delaware 19717-1303, USA.


Advances in Soil Science, Volume 6

VIII. (*con't*)
C. Rate of Exchange on Soils ..... 30
D. Use of Electroultrafiltration for Determining the Dynamics of K Exchange ..... 31
IX. Kinetics of Nonexchangeable and Mineral Potassium Release ..... 32
A. Dynamics of K Release from Micas ..... 32
B. Dynamics of Potassium Release from Feldspars ..... 36
C. Dynamics of Nonexchangeable Potassium Release from Soils ..... 39
D. Methods for Determining the Dynamics of Nonexchangeable Potassium ..... 40
X. Use of Kinetics for Determining Thermodynamics of Potassium Exchange in Soils ..... 44
A. Theoretical Considerations ..... 44
B. Experimental Results ..... 49
XI. Summary and Future Research Needs ..... 52
References ..... 52

# I. Introduction

The study of potassium (K) exchange on clay minerals and soils can be viewed from either a thermodynamic or a kinetic perspective. In pursuing the soil science literature, one will undoubtedly find that most of the studies on K exchange have been of an equilibrial rather than of a dynamic nature. Although the former studies have proved valuable in understanding the often elusive behavior of K in heterogeneous systems, they do not convey information about the rates or mechanisms of K exchange processes. Moreover, thermodynamic investigations on soil K are often not applicable to field conditions. Although a quasi-state equilibrium in K exchange can be achieved in the laboratory, these conditions are seldom if ever attained under field conditions.

Agricultural soils are nearly always in a state of disequilibrium with regard to K transformations. Soils that have been intensively cropped and fertilized with optimal K fertilizer additions for many years belong in this group because equilibrium is precluded by periodic additions of fertilizers (Sparks, 1985 a,b; Sparks and Huang, 1985). Goulding (1983), in an excellent review on thermodynamics of K exchange on clay minerals and soils, stated quite well the limitation in using thermodynamics to study K reactions under field conditions when he said: "When referring to the thermodynamics of K exchange, workers invariably mean exchange equilibria and thus equilibrium thermodynamics. The attainment of equilibrium in the laboratory is entirely possible, but in the field the exchange of K between soil and solution is a dynamic equilibrium, if it is an equilibrium at all."

G. W. Cooke (1979) has noted that one of the most important areas of agricultural research now and in the future is modeling of soil–plant processes. Certainly, cation exchange and in particular the dynamics of adsorption and release of K in soils must be an integral part of this research on modeling. To increase the efficiency of K fertilizer applications, and to predict the fate of applied K with time, one must have an understanding of the dynamic reactions between the phases of soil K. This type of predictive modeling would be particularly important on sandy soils of the world where substantial leaching occurs (Sparks, 1985a,b; Sparks and Huang, 1985).

J. Thomas Way (1850), whom many consider the patriarch of soil chemistry, was the first to study ion-exchange kinetics. In his classic and pioneering research, he found that the rate of $[NH_4]^+$ adsorption on a British soil was rapid and concluded that ion exchange was an instantaneous process. Further kinetic studies were not conducted until the 1900s, in the work of Gedroiz (1914) in Russia, Hissink (1924) in Holland, and Kelley (1948) in the United States. Gedroiz and Hissink also concluded that ion-exchange processes were very rapid. However, Professor W. P. Kelley (1948), in his beautiful treatise *Cation Exchange in Soils,* questioned whether these investigators' results would be true in all soils, as is so elegantly stated below:

> It is not difficult to see why the reaction should take place rapidly. The exchangeable cations, being located on the surface of layer silicates, the so-called inner crystalline swelling makes possible the rapid passage of the solution into the spaces between layer silicates where exchange takes place. It is doubtful whether exchange equilibrium is established as rapidly with the hydrous-mica-type of clay as with montmorillonitic clay, owing to the closeness of packing of the layer silicates of the former (pp. 51–52).

As we shall see later, Kelley was correct in stating that ion exchange might not always be an instantaneous process, particularly with K dynamics in vermiculitic and micaceous systems. Soil mineralogy is paramount in affecting the kinetics of K exchange on soils and clay minerals.

A complete discussion of thermodynamics of K exchange is not addressed in this contribution because these concepts have already been presented by Goulding (1983). Additionally, the work of Babcock (1963) and Sposito (1981) can be consulted for definitive discussions of equilibrial processes in soil systems. Neither is it the objective of this review to present in-depth discussions on pure chemical kinetics. For these, the reader should consult the excellent treatises by Frost and Pearson (1961), Laidler (1965), and Eyring *et al.* (1980).

Instead, the aim of this review is to examine thoroughly the dynamics of K reactions in soils and clay minerals and to apply the principles of chemical kinetics to these systems. Because most studies on kinetics of K reactions have been published only in the last 10 to 15 years, this contribution focuses on the recent scientific literature.

## II. General Characteristics of Soil K

Since the middle of the 17th century, when J. R. Glauker in The Netherlands first proposed that saltpeter ($KNO_3$) was the "principle" of vegetation, K has been recognized as being beneficial to plant growth (Russell, 1961). Glauker obtained large increases in plant growth from addition of saltpeter to the soil that was derived from the leaching of coral soils. The essentiality of K to plant growth has been known since the work of von Liebig published in 1840.

Of the major and secondary nutrient elements, K is usually the most abundant in soils (Reitemeier, 1951). Igneous rocks of the Earth's crust have higher K contents than sedimentary rocks. Of the igneous rocks, granites and syenites contain 46 to 54 g $kg^{-1}$ K, basalts contain 7 g $kg^{-1}$, and peridotites contain 2.0 g $kg^{-1}$. Among the sedimentary rocks, clayey shales contain 30 g of K per kilogram, whereas limestones have an average of only 6 g $kg^{-1}$ (Malavolta, 1985).

Mineral soils generally range between 0.04 and 3% K. Total K contents in soils range between 3000 and 100,000 kg $ha^{-1}$ in the upper 0.2 m of the soil profile. Of this total K content, ≈98% is bound in the mineral form whereas ≈2% is in soil solution and exchangeable phases (Schroeder, 1979; Bertsch and Thomas, 1985).

Some chemical characteristics of K and other common elements in the lithosphere are given in Table 1. Potassium, among mineral cations required by plants, is the largest in nonhydrated size ($r$ = 0.133 nm) and the number of oxygen ions surrounding it in mineral structures is high (8 or 12), which suggests that the strength of each K—O bond is relatively

**Table 1.** Chemical characteristics of K and other common elements in the lithosphere[a]

| Ion | Crystalline radii (nm) | | Hydrated radii (nm) | Polarizability ($nm^3$) | Debye-Hückel parameter (nm) | Coordination No. |
|---|---|---|---|---|---|---|
| $Li^+$ | 0.078 | 0.060 | 0.380 | 0.0079 | 0.432 | 6 |
| $Na^+$ | 0.098 | 0.095 | 0.358 | 0.0196 | 0.397 | 6, 8 |
| $K^+$ | 0.133 | 0.133 | 0.331 | 0.0876 | 0.363 | 8–12 |
| $Rb^+$ | 0.149 | 0.148 | 0.329 | 0.1407 | 0.349 | 8–12 |
| $Cs^+$ | 0.165 | 0.169 | 0.329 | 0.2452 | – | 12 |
| $Mg^{2+}$ | 0.078 | 0.065 | 0.428 | 0.0110 | 0.502 | 6 |
| $Ca^{2+}$ | 0.106 | 0.099 | 0.412 | 0.0523 | 0.473 | 6, 8 |
| $Sr^{2+}$ | 0.127 | 0.113 | 0.412 | 0.0880 | 0.461 | 8 |
| $Ba^{2+}$ | 0.143 | 0.135 | 0.404 | 0.1682 | 0.445 | 8–12 |
| $Al^{3+}$ | | 0.050 | | | | 4, 6 |
| $Si^{4+}$ | | 0.041 | | | | |

[a]From Gast (1977).

weak (Sparks and Huang, 1985). Potassium has a polarizability equal to 0.088 $nm^3$, which is higher than for $Ca^{2+}$, $Li^+$, $Mg^{2+}$, and $Na^+$ ions but lower than for $Ba^{2+}$, $Cs^+$, $[NH_4]^+$, and $Rb^+$ (Rich, 1968, 1972; Sparks, 1980; Sparks and Huang, 1985). Ions with higher polarizability are preferred in ion-exchange reactions. Potassium has a hydration energy of 142.5 kJ $g^{-1}$ $ion^{-1}$, which indicates little ability to cause soil swelling (Helfferich, 1962).

# III. Forms of Soil Potassium

Dynamic equilibrial reactions exist between the phases of soil K (Figure 1). The forms of K in soil in the order of their availability to plants and microbes are solution, exchangeable, nonexchangeable, and mineral (Martin and Sparks, 1985; Sparks and Huang, 1985). All of these forms may be quantified as shown in Table 2.

Soil solution K is the form taken up directly by plants and microbes and is also subject to leaching (Sparks, 1980). It is usually found in low quantities. The concentration of K in the soil solution is enigmatic. It fluctuates greatly and is difficult to measure. Because the soil solution is polyionic and is often fairly concentrated, the thermodynamic activity, rather than just the molar or molal concentration of K, should be determined if a picture of what the plant root "sees" is desired (Adams, 1974). Levels of soil solution K are determined by the equilibria and kinetic reactions between the other forms of soil K, the soil moisture content, and

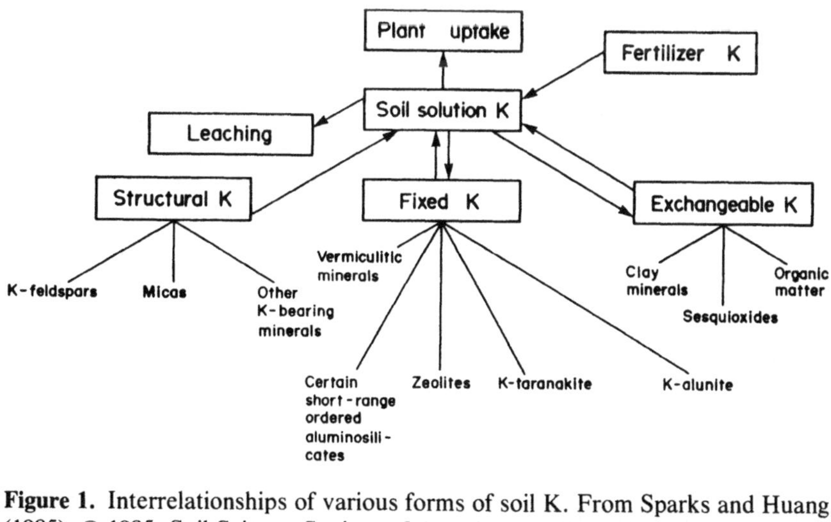


**Figure 1.** Interrelationships of various forms of soil K. From Sparks and Huang, (1985). © 1985, Soil Science Society of America, Madison, Wisconsin.

**Table 2.** Forms of soil K and extraction methods that are commonly used in K analysis of clays and soils[a]

| Form | Location | Extractants |
|---|---|---|
| Water soluble | Soil solution | Column displacement<br>Pressure membrane<br>Immiscible displacement and centrifugation |
| Exchangeable | Colloidal exchange sites— clay and organic matter | $NH_4OA_c$<br>$N$ $NH_4Cl$,<br>Dilute $H_2SO_4$ and HCl,<br>$N$ $CaCl_2$ or $MgCl_2$<br>Dilute $CaCl_2$ or $MgCl_2$<br>Electrodialysis<br>Electroultrafiltration<br>Silver thiourea |
| Nonexchangeable | Vermiculite<br>Trioctahedral mica<br>Dioctahedral mica<br>Hydrous mica (Illite)<br>Chlorite–vermiculite intergrades<br>Interstratified mica–smectites<br>x-ray amorphous minerals | Exhaustive cropping<br>Exhaustive leaching with $0.01N$ HCl<br>Equilibration with $0.5N$ HCl<br>Strong HCl at 373K<br>Boiling 23% HCl<br>Exhaustive leaching with $0.1N$ NaCl<br>Sodium Cobaltinitrate<br>Hot $MgCl_2$<br>Successive moist incubations and salt leachings<br>Equilibration with sodium tetraphenylboron ($NaBPh_4$)<br>Serial extractions with $NaBPh_4$<br>Boiling $HNO_3$<br>Electrodialysis<br>Electroultrafiltration<br>Serial extractions with Ca-saturated cation-exchange resin<br>Equilibration with H-saturated cation-exchange resin |
| Mineral | Trioctahedral mica<br>Dioctahedral mica<br>Orthoclase (K-feldspar) | Selective dissolution with Na-pyrosulfate fusion |
| Total | — | HF digestion |

[a]Adapted from Sparks (1980).

the divalent ion content in solution and on the exchanger phase (Sparks and Huang, 1985).

Exchangeable K is held by the negative charges of organic matter and clay minerals. It is easily exchanged with other cations and is readily available to plants. The release of exchangeable K to the soil solution is called desorption, whereas the reverse reaction is termed adsorption.

Nonexchangeable K is distinct from mineral K in that it is not bonded covalently within the crystal structures of soil mineral particles. Instead, it is held between adjacent tetrahedral layers of dioctahedral and trioctahedral micas, vermiculites, and intergrade clay minerals (Rich, 1972; Sparks and Huang, 1985). If nonexchangeable K is equated to "fixed" K, then it can also occur in random gaps in the structure of x-ray amorphous clay-sized minerals (Barber, 1979). Nonexchangeable $K^+$ ions held in these interlayers and gaps are bound coulombically to the negatively charged interlayer surface sites (Kittrick, 1966). This binding force exceeds the hydration forces between individual $K^+$ ions, resulting in a partial collapse of the crystal structure (Kittrick, 1966). Therefore, the $K^+$ ions are physically trapped to varying degrees, making diffusion the rate-limiting step for K release (Martin and Sparks, 1983).

Nonexchangeable K can also be found in "wedge zones" of weathered micas and vermiculites (Rich, 1964). These "wedge zones" are too narrow for exchanging $Ca^{2+}$ or $Mg^{2+}$ ions to enter; however, $[NH_4]^+$ and $[H_3O]^+$ ions, because of their similar hydrated radii, can enter these zones (Rich, 1964).

Nonexchangeable K is moderately to sparely available to plants, depending on various soil parameters (Goulding and Talibudeen, 1979; Sparks and Huang, 1985). Release of nonexchangeable K to the exchangeable form occurs when levels of exchangeable and soil solution K are decreased (Doll and Lucas, 1973) by crop removal and/or leaching (Sparks *et al.*, 1980a) and perhaps by large increases in microbial activity.

The bulk of the total K in most soils is in the mineral form (Sparks and Huang, 1985). An example of this is shown for an Atlantic Coastal Plain soil from Delaware (Table 3). In all horizons of this soil, exchangeable and nonexchangeable K constitute a very small percentage of the total K. The bulk of the total K is in the mineral form as feldspars.

Mineral K is generally assumed to be only slowly available to plants (Jackson, 1964; Sparks and Huang, 1985); however, the availability is dependent on a number of factors, including the level of K in the other phases, viz., solution, exchangeable, and nonexchangeable, and the degree of weathering of the feldspars and micas making up the mineral K fraction. Common soil K-bearing minerals in the order of availability of their K to plants are biotite, muscovite, orthoclase, and microcline (Huang *et al.*, 1968). It has generally been thought that only small amounts of feldspar and mica K are released over a growing season (Rasmussen, 1972) but recently, Sparks and his co-workers (Mirecki *et al.*, 1985; Sadusky and

Table 3. Distribution of potassium in a Delaware soil[a]

| Depth (m) | Horizon | Exchangeable K (cmol kg$^{-1}$) | Nonexchangeable K (cmol kg$^{-1}$) | Mineral[b] K (cmol kg$^{-1}$) | Total K (cmol kg$^{-1}$) |
|---|---|---|---|---|---|
| 0–0.09 | Ap | 0.25 | 0.42 | 35.0 | 35.7 |
| .09–.23 | E | 0.18 | 0.40 | 42.8 | 43.4 |
| .23–.28 | EB | 0.24 | 0.48 | 41.7 | 42.4 |
| .28–.34 | Bt1 | 0.26 | 0.47 | 43.7 | 44.4 |
| .34–.47 | Bt2 | 0.25 | 0.49 | 45.3 | 46.0 |
| .47–.56 | Bt3 | 0.24 | 0.41 | 46.9 | 47.5 |
| .56 + | C | 0.21 | 0.37 | 47.3 | 47.9 |

[a]From Hendricks *et al.* (1986).
[b]Estimated by subtracting the sum of exchangeable and nonexchangeable K from total K contents.

Sparks, 1985) have found that a substantial amount of K is being released from the sand fractions of Delaware soils. The sand mineralogy of these soils is high in feldspars that are very weathered. The K release from the feldspars could explain the anomalous crop response behavior a number of researchers have observed with corn (*Zea mays* L.) on Atlantic Coastal Plain soils (Yuan *et al.*, 1976; Sparks *et al.*, 1980a; Hendricks *et al.*, 1985). The release of mineral K to more available forms is referred to as weathering or, in severe cases, as dissolution. The reverse of this reaction is immobilization or precipitation.

## IV. Inorganic and Organic Colloids Important in Potassium Dynamics

The importance of organic and particularly inorganic components in controlling the dynamics of soil K cannot be overemphasized. Indeed, fully to comprehend and explain the complex and often enigmatic behavior of K in soils, one must have a thorough knowledge of the mineralogy. An attempt will be made here to describe briefly the structure and occurrence of inorganic and organic colloids important in K dynamics. Table 4 lists the formulas and some basic properties for the inorganic colloids. For more complete discussions, one should consult the work of Rich (1968, 1972), Rasmussen (1972), von Reichenbach (1972), Stitcher (1972), Schroeder (1974, 1979), Dixon and Weed (1977), and Sparks and Huang (1985).

### A. Micas

The mica structure consists of negatively charged 2:1 layers that are bound together by large interlayer cations. In the K-bearing micas, the in-

**Table 4.** Characteristics of inorganic soil components important in potassium dynamics[a]

| Mineral | Chemical Formula | CEC (cmol kg$^{-1}$) | Surface area (m$^2$ g$^{-1}$) | Potassium content (g kg$^{-1}$) |
|---|---|---|---|---|
| Kaolinite | $Si_4Al_4O_{10}(OH)_8$ | 2–15 | 7–30 | — |
| Dioctahedral smectite | $M^+_{.66}H_2O\ Al_{3.34}(Fe^{2+}, Mg)_{.66}Si_8O_{20}(OH)_4$ | 80–150 | 600–800 | — |
| Dioctahedral vermiculite | $M_{1.48}H_2OAl_{2.8}M_{0.60}Fe^{3+}_{0.6}Si_{7.12}Al_{.88}O_{20}(OH)_4$ | 10–150 | 50–800 | |
| Muscovite | $K_2Al_4Si_6Al_2O_{20}(OH,F)_4$ | 10–40 | 60–100 | 80 |
| Biotite | $K_2(MgFe,Al)_6(Si,Al)_8O_{20}(OH,F)_4$ | 10–40 | 40–100 | 70 |
| Allophane | $SiO_2\ Al_2O\ 2H_2O$<br>$2\ SiO_2Al_2O_3\ 3H_2O$ | 5–350 | 100–800 | 1.0–1.3 |
| Feldspars (general) | $KAlSi_3O_8$ | – | — | |
| Orthoclase | $(K,Na)\ Al\ Si_3O_8$ | | | ~110 |
| Sanidine | $K\ Al\ Si_2O_6$ | | | |
| Microcline | $(Na, K)\ Al\ SiO_4$ | | | |

[a] Partially adapted from Malavolta (1985).

terlayer cation is mainly the $K^+$ ion. The 2:1 layer is composed of an octahedron sheet between two sheets of tetrahedra. In trioctahedral micas, such as biotite, all three octahedral positions are filled, whereas in dioctahedral micas, such as muscovite, only two out of three octahedral cation positions are filled (Fanning and Keramidas, 1977; Sparks and Huang, 1985).

The layer charge in K-bearing micas ideally is $-1.0$ per formula unit and arises by some combination of three mechanisms: (1) substitution of $R^{3+}$ (primarily Al, Fe, or Cr) for $Si^{4+}$ in tetrahedral positions, (2) substitution of $R^+$ or $R^{2+}$ for $R^{2+}$ or $R^{3+}$ in octahedral positions, or (3) vacancies in octahedral positions. The resultant layer charge may originate entirely within the tetrahedral sheet, may originate entirely within the octahedral sheet in some species, or may come partly from both sheets. When the tetrahedral charge exceeds $-1.0$, the octahedral sheet possesses an excess positive charge.

Micas are the third most extensive group of minerals after feldspars and quartz in granite and sialic (acid) rocks in general but are less extensive in most mafic rocks (Holmes, 1965). Muscovite and biotite are the most extensive micas in igneous and metamorphic rocks. Phlogopite occurs as a product of metamorphism of magnesian limestone or dolomitic limestones and also in serpentinitic rocks. Micas are generally more extensive in fine-grained sediments and sedimentary rocks (clays, shales) than in coarser textured seimentary rocks (e.g., sandstones). Shales and slates are usually rich in the fine-grained illitic-type micas. Illitic micas are also important clay minerals in limestones.

In most soils, micas originate mainly from soil parent materials and tend to weather to other minerals with time. Therefore, they are in general more prevalent in younger and less weathered soils (Entisols, Inceptisols, Mollisols, Andosols, Alfisols) and less prevalent in more weathered soils (Ultisols, Oxisols) (Jackson *et al.*, 1952; Jackson, 1964).

In soils that are highly weathered, trioctahedral micas are uncommon. Therefore, the micas in clays of most soils are predominantly dioctahedral.

## B. Feldspars

The structure of K-feldspars is a three-dimensional framework of linked $SiO_4$ and $Al_2O_3$ tetrahedra, with sufficient opening in the framework to accommodate K to maintain electroneutrality (J. V. Smith, 1974; Sparks and Huang, 1985). In building up the framework, four-membered rings of tetrahedra are the basic units. These rings are linked together to form a honeycomb type of arrangement.

The K-feldspar polymorphs—sanidine, orthoclase, microcline, and adularia—have identical chemical compositions (Barth, 1969; Sparks and Huang, 1985). Sanidine is a monoclinic alkali feldspar with small optic axial angle (2V) and commonly occurs in volcanic rocks. The monoclinic alkali feldspars, which have the larger optic axial angle (2V), look homogeneous, and do not show cross-hatched twinning, are referred to as orthoclase. Microcline is triclinic, exhibits the typical cross-hatched twinning, and has the larger optical axial angle (2V). The alkali feldspars, which may be either monoclinic or triclinic but have special habits and occur in low-temperature hydrothermal veins, are referred to as adularia.

In the K-feldspar structure, one out of every four Si atoms in the framework is replaced by Al. This substitution imparts a negative charge to the framework, which is neutralized by the incorporation of other positively charged ions, such as $K^+$. Only rarely do K-feldspars of ideal chemical composition ($KAlSi_3O_8$) occur in nature. The K-feldspars contain foreign cations either in fourfold coordination, replacing Al or Si, or in higher coordination, replacing $K^+$ ions. Lithium and Be have the suitable size relationship to occupy the tetrahedral positions (Heier, 1962). Frequently, Na substitutes for K in the K-feldspar structure, and a complete solid solution series exists between the end members of composition $KAlSi_3O_8$ and $NaAlSi_3O_8$. The K- and Na-rich members of the group with a small amount of Ca are known as alkali feldspars. The lamellar aggregate, which is composed of a large amount of alkali feldspar, a subordinate amount of alkali feldspar, and a subordinate amount of albite ($NaAlSi_3O_8$), is referred to as perthite. Most naturally occurring alkali feldspars, except authigenic K-feldspars, usually contain varying amounts of Na in the structure and are therefore more or less perthitic (Barth, 1969). The other ions that may occupy K positions are Rb, Cs, Pb, Tl, Ca, Si, Cu, Ga, and Ba (Barth, 1969).

The polymorphs of K-feldspars make up 16% of the total earth's crust (Ahrens, 1965; Rasmussen, 1972). When the alkali feldspars that contain K are taken into account, the total K-bearing feldspars make up nearly 31% (Barth, 1969). Microcline is the common K-feldspar of pegmatites and hydrothermal veins and also occurs in metamorphic rocks. Orthoclase is the characteristic K-feldspar of igneous rocks; it occurs both alone and in perthitic intergrowth with albite. Orthoclase also occurs in metamorphic rocks. Sanidine is present in K-rich volcanic rocks, such as rhyolite and trachyte. Adularia occurs in low-temperature hydrothermal veins.

A major portion of K-feldspars found in sedimentary rocks is of igneous origin. They accumulate in sedimentary environments as a weathering residue of the igneous and metamorphic rocks (Rankama and Sahama, 1950).

The quantities of K-bearing feldspars in soils vary with the intensity and capacity factors of weathering reactions. In strongly weathered soils, K-feldspars are present in only small quantities or are completely absent, although the parent material contains considerable quantities of those minerals (van der Plas, 1966). However, some feldspars may be found in humid tropical soils, which contain relatively fresh rock materials because of erosional and depositional processes (Jackson and Sherman, 1953).

Feldspars are commonly present in the silt and sand fractions of the young to moderately developed soils representing various soil parent materials and soil-forming conditions (Somasiri *et al.*, 1971; Sparks and Huang, 1985). Alkali feldspars are present even in the clay fraction of soils formed under moderate weathering (Huang and Lee, 1969).

The K-feldspars and K-micas of soils serve as an important reserve of available $K^+$. The distribution of K in feldspars and micas in a series of particle-size fractions of moderately weathered soil profiles is shown in the work of Somasiri *et al.* (1971) (Table 5). The fraction of K from feldspars increases with increasing particle size. The opposite trend is true for the fraction of K from micas. Sadusky and Sparks (1985) have also found considerable quantities of feldspars in sandy, Atlantic Coastal Plain soils of Delaware.

### C. Vermiculite

Vermiculite has a layer charge of usually $> -0.6$ equivalents per formula unit (Bailey, 1980). The structure of vermiculite consists of 2 : 1 layers and exchangeable cations that have a plane of water molecules on each side. Water molecules are tetrahedrally bonded to oxygens of the 2 : 1 layer surfaces through hydrogen bonding (Sparks and Huang, 1985).

Vermiculites can be found in all of the major soil great groups. However, they are found more often in soils of temperate and subtropical climates than in soils of tropical areas. Vermiculites are present in all particle-size fractions ranging from clay through coarse sand.

Soil vermiculites are nearly always reported to accompany muscovite, biotite, and chlorite. It is generally believed that all vermiculites are alteration products of micas or chlorites (Douglas, 1977). Vermiculites can

**Table 5.** Distribution of feldspar K in various size fractions of a Haverhill Ap soil horizon (Aridic Haploboroll) in Saskatchewan, Canada[a]

| Particle size (μm) | Total K ($g\ kg^{-1}$) | Feldspar K ($g\ kg^{-1}$) | Mica K ($g\ kg^{-1}$) | % fraction of K Feldspar | % fraction of K Mica |
|---|---|---|---|---|---|
| <0.2 | 21.1 | — | 21.1 | — | 100 |
| 0.2–2 | 26.1 | 0.8 | 5.3 | 3 | 97 |
| 2–5 | 20.1 | 6.1 | 14.0 | 30 | 70 |
| 5–20 | 17.3 | 8.7 | 8.6 | 50 | 50 |
| 20–50 | 15.4 | 9.7 | 5.7 | 63 | 37 |
| 50–500 | 12.7 | 7.6 | 5.1 | 60 | 40 |
| 500–2000 | 17.3 | 17.3 | — | 100 | — |
| <2000+ | 16.9 | 6.8 | 10.1 | 40 | 60 |

[a]Somasiri *et al.* (1971).
Permission granted from Soil Science Society of America, Madison, Wisconsin.

form from a variety of parent materials under many different soil environments.

Dioctahedral vermiculite is more common than trioctahedral vermiculite in soils (Jackson, 1959; Rich, 1960). Trioctahedral vermiculite may comprise a significant amount of the sand and silt fractions of soils (Alexiades *et al.*, 1973; Coffman and Fanning, 1974), whereas dioctahedral vermiculite is seldom observed as discrete crystallite >5 μm (Douglas, 1977). Medium-sized silt fractions may contain muscovite–dioctahedral vermiculite interstratified minerals. However, fine-grained vermiculite in soils and sediments may be either trioctahedral or dioctahedral (Bailey, 1980).

As the particle size of soils decreases, the ratio of dioctahedral vermiculite to muscovite increases (Rich, 1958). In many soils, the ratio of vermiculite to mica decreases from the soil surface downward (Douglas, 1977). The transformation of mica to vermiculite depends on a strong leaching environment to remove solution K. The intensity of the weathering reaction decreases with increasing depth in soil profiles.

In intensive weathering environments, vermiculites invariably have hydroxy-Al interlayers (Rich, 1958). Hydroxy-Al-interlayered vermiculites may be either dioctahedral (Rich and Obenshain, 1955) or trioctahedral (Kato, 1965). Hydroxy-Al-interlayered dioctahedral vermiculite can persist under intense weathering conditions (Sparks and Huang, 1985).

## D. Allophane

Allophane is not a distinct mineral species but a series of minerals that are hydrous aluminosilicates of widely varying chemical composition, characterized by short-range order, the presence of Al—O—Si bands, and a differential thermal analysis (DTA) curve displaying a low-temperature endotherm and a high-temperature exotherm without an intermediate endotherm (van Olphen, 1971; Wada, 1977; Sparks and Huang, 1985). Allophane has a definite affinity for K because of its structural configuration. Wada and Harada (1969) obtained unusually high cation-exchange capacity (CEC) values for Japanese soil allophanes when determined with K salts. The fixation of K by synthetic aluminosilica gels is sterically controlled, and channels in the gels impose a sieve action on the entry and passage of certain counterions (van Reeuwijk and de Villiers, 1968).

## E. Zeolites

Zeolites are a large group of hydrated aluminosilicates consisting of a tetrahedral framework of O atoms surrounding either a Si or an Al atom extended in a three-dimensional network that provides structural channels. Each O atom is shared between two tetrahedra. The net negative charge arising from the isomorphous substitution of $Al^{3+}$ for $Si^{4+}$ is balanced by the presence of alkali and alkaline earth cations, most commonly $Ca^{2+}$, $Na^{+}$, and $K^{+}$, within the existing pores.

Ion-exchange experiments have shown that zeolites have a stronger affinity for $K^+$ than for $Na^+$, $Ca^{2+}$, and $Mg^{2+}$ (Ames, 1960; Barrer *et al.*, 1969; Barrer and Munday, 1971). Potassium ions may occupy positions in the structural channel of zeolites from which $K^+$ ions cannot be removed without destroying the crystal structure (Peterson *et al.*, 1965).

### F. Smectites

The smectites include all minerals formerly classified in the montmorillonite group of expansible layer silicates. In soil environments, important smectites are montmorillonite, beidellite, and nontorite. The structure of smectite resembles that of mica and vermiculite in that all three have an octahedral sheet in coordination with two tetrahedral sheets in which oxygen atoms are shared. Cationic substitution occurs in octahedral or tetrahedral sheets (Borchardt, 1977; Sparks and Huang, 1985). Smectites, along with vermiculite, are responsible for a large portion of the CEC of soils in temperate regions. A commonly accepted definition of smectite specifically rules out minerals that fixed K (Alexiades and Jackson, 1966). However, smectites are likely to have residual mica cores surrounded by "wedge sites" that may be primarily responsible for K fixation (Rich, 1964; Rich and Black, 1964).

### G. Chlorites and Hydroxy-Interlayered Expansible Phyllosilicates

Chlorites in soils are largely inherited as primary minerals found in metamorphic or igneous rocks or occur as alteration products from such minerals as hornblende, biotite, and other ferromagnesian minerals. The abundance and frequency of occurrence of chlorites in soils are relatively low. The CEC of chlorite is also relatively small (Barnhisel, 1977; Sparks and Huang, 1985).

Hydroxy-interlayered vermiculite and smectite occur in soils as a product of weathering (Rich, 1968, 1972). They may result from degradation of chlorite or deposition of hydroxy materials within the interlayer spaces of expansible layer silicates. The geographical distribution of hydroxy-interlayered minerals is wide and they may occur in the soils of several orders of the soil classification system (Soil Survey Staff, 1975). Their frequency of occurrence is highest in the Ultisols and the Alfisols. The CEC of hydroxy-interlayered vermiculites and smectites has a wide range depending on the degree of filling of the interlayer space. Formation of hydroxy interlayers results in the reduction of the CEC of clay minerals (Rich and Obenshain, 1955; Rich, 1964; Barnhisel and Bertsch, 1982) and they have pronounced effects on K selectivity of vermiculite (Rich and Black, 1964; Kozak and Huang, 1971; Sparks and Liebhardt, 1982; Sparks and Huang, 1985).

### H. Kaolinite and Halloysite

Kaolinite is a widespread clay mineral in soils. It is most abundant in soils of warm moist climates. Halloysite is formed rapidly in soils of volcanic origin. Kaolinite and halloysite are products of acid weathering. They are 1 : 1 layer structural aluminosilicates of the same ideal composition, except that in hydrated halloysites there is water between the layers.

Permanent negative charge on kaolinite from isomorphous substitution has been postulated because exchangeable cations are retained under acid conditions (Schofield and Samson, 1954). More recent findings have contributed to understanding isomorphous substitution and the CEC of kaolinite. Smectite interstratification with kaolinite and mica zones in kaolinite particles have been reported (Atschuler *et al.*, 1963; Lee *et al.*, 1975a,b).

Relatively few "pure" kaolins exist in nature and soils (Komarneni, 1978). This has recently been documented in studies of Talibudeen and Goulding (1983). Through microcalorimetry studies, they showed that the differential heats of $K^+$–$Ca^{2+}$ exchange exhibit that six groups of sites (ranging from $-13.8$ to $-5.1$ kJ $mol^{-1}$ $K^{-1}$ exchanged and with as many as four in any one sample) exist in kaolins. These heat values, coupled with entropies of exchange, suggest that 0.1 to 10% vermiculitic, micaceous, and smectitic layers are present in kaolins. Therefore, the permanent negative charge attributed to kaolinite probably arises from these impurities.

A number of researchers have shown that kaolins show an affinity for $K^+$ over $Ca^{2+}$ (H. E. Jensen, 1973; Udo, 1978). However, this is probably a result of impurities in the kaolins, such as vermiculite and micas, which exhibit a pronounced affinity for $K^+$ over $Ca^{2+}$ (Jardine and Sparks, 1984a,b). This hypothesis is substantiated by a recent study in which Carski and Sparks (1985) studied K–Ca exchange on a relatively pure kaolin and found $Ca^{2+}$ was indeed preferred over $K^+$.

### I. Potassium-Taranakite

Potassium-taranakite is formed in soils through the reaction of concentrated K and acid phosphate solutions with Al hydroxides (Taylor and Gurney, 1965). This compound can be hydrolyzed (Taylor *et al.*, 1963) upon dilution and is logically considered to be a slow-releasing K source.

### J. Potassium-Alunite

Potassium-alunite can be formed in acid soils if the ionic activities of $K^+$, $Al^{3+}$, and $[SO_4]^{2-}$ are sufficiently high. The negative logarithm of the solubility product ($pK_{sp}$) of alunite is 85.40 (Adams and Rawajfih, 1977). The solubility data indicate the feasibility of simultaneous retention of $K^+$ and

$[SO_4]^{2-}$ in the form of alunite in acid soil environments (Sparks and Huang, 1985).

### K. Organic Matter

Organic colloids possess negative charges that arise from the dissociation of carboxyl and phenolic hydroxyl groups (Huang, 1980; Sparks and Huang, 1985). These groups certainly adsorb $K^+$ ions from soil solution. However, the number of exact measurements of K-exchange equilibria in well-defined systems, wherein the adsorbent is organic colloids, is very limited and merits further research (Schuffelen, 1972).

Previous research, however, does indicate that organic matter exhibits a preference for divalent cations, such as $Ca^{2+}$ and $Mg^{2+}$, over $K^+$ (Naylor and Overstreet, 1969). Jardine and Sparks (1984b), investigating K–Ca exchange in a multireactive soil system, found that $Ca^{2+}$ was preferred to $K^+$ on organic matter sites.

## V. Techniques for Measuring Potassium Dynamics

One of the most important aspects of any kinetic study with K is the method one employs to measure rate parameters. Soil chemists have used primarily batch and miscible displacement or flow techniques to investigate kinetics of K reactions in soils and clay minerals (Sparks, 1985a,b).

### A. Batch Techniques

The bulk of kinetic studies with soils and clay minerals have employed batch techniques. These usually involve placing an exchanger and the adsorbate in a vessel, such as a centrifuge tube. The suspension is then agitated using a reciprocal shaker, or it is stirred. Then, the suspension is usually centrifuged to obtain a clear supernatant solution for subsequent analysis.

The use of batch techniques to measure the rate of K reactions in clay minerals and soils has a number of disadvantages. Unless the technique of Zasoski and Burau (1978) is used, most batch techniques require centrifugation, which requires about 5 min to separate the solid from the liquid phases. Many exchange reactions are complete in this time or less (Harter and Lehmann, 1983; Jardine and Sparks, 1984a; Sparks, 1985a,b; Sparks and Huang, 1985).

Also, to measure the kinetics of a chemical reaction properly, the technique should not change the reactant concentration (Zasoski and Burau, 1978). Therefore, the sample and the suspension should have a similar solid to solution ratio at all times. Unfortunately, this has not been true in most batch studies, as Barrow (1983) discusses in an excellent review article. Most studies have employed large solution/soil ratios where the concentration in the solution and the quantity of adsorption both vary simultaneously. Exceptions are the studies by Zasoski and Burau (1978) and Van Riemsdjik and deHaan (1981). In the latter two studies, the solution concentration was held constant. Thus, the relation between adsorption and time should only be treated as a simple, two-dimensional relation if adsorption is constant. Unless a batch technique similar to that of Zasoski and Burau (1978) and Van Riemsdjik and deHaan (1981) is utilized, these conditions do not apply to experiments in which a wide solution/soil ratio is used.

### B. Flow or Miscible Displacement Techniques

Flow or miscible displacement techniques have been used to a lesser extent to investigate the dynamics of K reactions in clay minerals and soils (Sivasubramaniam and Talibudeen, 1972; Sparks *et al.*, 1980c; Sparks and Jardine, 1981; Jardine and Sparks, 1984a; Sparks, 1985a,b). However, flow techniques may be preferable to batch studies in examining K sorption–desorption phenomena on colloids, particularly if one wishes to relate kinetic studies to solute transport under field conditions (Murali and Aylmore, 1981, 1983a,b). Sparks *et al.* (1980c) developed a miscible displacement technique to study kinetics of K adsorption on soils. Later, Jardine and Sparks (1984a) modified the technique so that aliquots of leachate could be taken at 2-min increments. Thus, one could investigate very rapid ionic reactions that could not be measured with most batch techniques.

### C. Critical Evaluation of Kinetic Techniques

Recently, Ogwada and Sparks (1986a,b) investigated the effect of kinetic methodology on K adsorption rate coefficients ($k_a$) for K adsorption using five different methods (Table 6). These methods were a miscible displacement, batch, static, stirred, and vortex batch. Kinetics of $K^+$ adsorption were studied on a Ca-saturated kaolinite, a Chester soil (fine-loamy, mixed mesic Typic Hapludults) and vermiculite.

The magnitude of the $k_a$ values for the three adsorbents studied by Ogwada and Sparks (1986b) was clearly affected by the kinetic method employed (Table 6). The type of method also greatly affected the time required for equilibrium in K adsorption to be reached in the three systems. For example, with kaolinite, 60 min were required in the static system without any agitation, whereas only 0.58 min was required with the vortex batch method. The authors observed similar trends with the Chester soil and vermiculite. The time required for equilibrium to be reached in the three systems followed the order: static > miscible displacement > batch > stirred > vortex batch.

Many workers have assumed that batch techniques greatly reduce or

**Table 6.** Effect of kinetic method on adsorption rate coefficients ($k_a$) in systems studied[a]

| | | Kinetic method ($k_a$, $min^{-1}$) | | | | |
|---|---|---|---|---|---|---|
| | Temperature | Static | Miscible displacement | Batch | Stirred | Vortex batch |
| Kaolinite | 283 | 0.030 | 0.033 | 1.304 | 1.321 | 1.444 |
| | 298 | 0.034 | 0.037 | 2.383 | 2.838 | 3.482 |
| | 313 | 0.037 | 0.045 | 3.764 | 3.784 | 4.981 |
| Chester loam | 283 | 0.025 | 0.031 | 0.179 | 0.192 | 0.867 |
| | 298 | 0.027 | 0.036 | 0.225 | 0.250 | 2.331 |
| | 313 | 0.031 | 0.042 | 0.260 | 0.293 | 3.240 |
| Vermiculite | 283 | 0.021 | 0.012 | 0.058 | 0.049 | 0.421 |
| | 298 | 0.024 | 0.015 | 0.083 | 0.056 | 0.945 |
| | 313 | 0.027 | 0.018 | 0.099 | 0.069 | 1.647 |

[a]From Ogwada and Sparks (1986b).

eliminate the effect of diffusion in ion-exchange processes. However, the work of Ogwada and Sparks (1986a,b) indicates that in systems where intraparticle diffusion can occur, such as with K adsorption on vermiculite, the batch technique is not successful in eliminating the effect of diffusion.

The question then becomes which technique should be used to study the dynamics of K reactions in soils and clay minerals. Both batch and miscible displacement techniques have advantages and disadvantages. With the batch technique shaking or stirring is usually involved. If mixing is inadequate, the rate of reaction becomes diffusion limited. However, vigorous mixing, which many investigators have used, can cause abrasion of the exchanger particles, leading to high rates of reaction and alterations in the surface chemistry of the particles (Barrow and Shaw, 1979). This problem could be serious when studying K dynamics because release could occur from K-bearing minerals.

The miscible displacement or flow technique is not a panacea solution to studying K dynamics. However, it appears to have a number of benefits. As mentioned earlier, rapid exchange reactions can be observed. Additionally, with solution flowing through the soil system, the solution not only brings in more ions, but also removes the desorbed ions of other species that have been present originally at potential sorption sites. This is particularly important in studying K reactions because small amounts of K in the equilibrium solution prevent further release of adsorbed K, which, consequently, results in marked hysteresis (Martin and Sparks, 1983). With a miscible displacement technique, one can obtain almost complete reversibility in K exchange because desorbed $K^+$ ions are constantly being removed.

However, along with the advantages of a flow technique for studying K dynamics, one must also recognize some disadvantages. Dilution errors can exist with the original miscible displacement technique of Sparks *et al.* (1980c) which could cause errors, particularly in systems that are known to adsorb very small quantities of ions, such as with B sorption on sandy soils (Evans, 1983; Carski and Sparks, 1985). Fortunately, in these systems, the dilution problem can be eliminated by using a modified miscible displacement technique developed by Carski and Sparks (1985).

Another disadvantage of a miscible displacement technique is the pronounced diffusion-controlled exchange that occurs; this presents problems if one is attempting to use kinetics to calculate thermodynamic parameters. Recently, Ogwada and Sparks (1986a) have shown that if diffusion is rate limiting, kinetics cannot be used to determine thermodynamic parameters.

# VI. Equations and Models to Describe Potassium Dynamics in Soils and Clay Minerals

The application of chemical kinetics to soil or clay mineral systems is indeed a challenging and difficult task. This is in large part because of the heterogeneous nature of both systems. However, the information one can gather from such an application is fundamental to understanding the rates and mechanisms of ionic reactions.

## A. Basic Equations

A number of equations have been employed to describe the kinetics of K reactions in soils and clay minerals (Sparks and Jardine, 1984; Havlin and Westfall, 1985). These have included the first-order, Elovich, parabolic diffusion, zero-order, and power-function equations.

These equations have been used to determine reaction order and rate coefficients between the four phases of soil K. The data gained from the application of these equations have proved useful in better understanding the mechanisms and rates of K adsorption and desorption in soils.

### 1. First-Order Equation

A number of researchers have successfully used a simple first-order equation to describe K exchange on clay minerals and soils (Sivasubramaniam and Talibudeen, 1972; Sparks *et al.*, 1980b; Sparks and Jardine, 1981; Jardine and Sparks, 1984a; Sparks and Jardine, 1984); nonexchangeable K release from soils (Burns and Barber, 1961; Martin and Sparks, 1983; Sadusky and Sparks, 1985); and K release from K-bearing minerals (Mortland, 1958; Mortland and Ellis, 1959; Mortland, 1961; Reed and Scott, 1962; Huang *et al.*, 1968; Scott, 1968; Feigenbaum and Shainberg, 1975; Mirecki *et al.*, 1985). The final forms of first-order equations to describe K adsorption and desorption processes using batch and miscible displacement techniques are given in Table 7.

**Table 7.** First-order equations for describing potassium adsorption and desorption processes using batch and miscible displacement methods

| Process | Kinetic method: Batch | Kinetic method: Miscible Displacement |
|---|---|---|
| Adsorption | $\ln K = \ln K_0 - k_a t$ [a] | $\ln (1 - K_t/K_\infty) = k_a K_\infty t$ [b] |
| Desorption | $\ln (K_0 - K) = \ln K_0 - k_d t$ [c] | $\ln (K_t/K_0) = k_d t$ [d] |

[a] $K$ = concentration of K in solution at time $t$; $K_0$ = initial concentration of K added at time zero; $k_a$ = adsorption rate coefficient.

[b] $K_t$ = amount of K on the colloid at time $t$; $K_\infty$ = amount of K on the colloid at equilibrium.

[c] $K_o$ = total amount of K that could be released at equilibrium; $K$ = amount of K released at time $t$; $k_d$ = desorption rate coefficient.

[d] $K_o$ = amount of K on the exchange sites of the colloid at zero time of desorption; $K_t$ = amount of K on the exchange sites of the colloid at time $t$.

Sparks and Jardine (1984) studied the kinetics of K adsorption on the standard clay minerals kaolinite, montmorillonite, and vermiculite. The first-order equation described K adsorption on kaolinite and montmorillonite well and two simultaneous first-order reactions appeared to describe K adsorption on vermiculite (Figure 2).

Jardine and Sparks (1984a) found that K adsorption (Figure 3) on an Evesboro soil (mesic, coated Typic Quartzipsamment) from Delaware conformed to first-order kinetics at 283, 298, and 313K. At 283 and 298K, two simultaneous first-order reactions existed, with the first slope containing both a rapid reaction (reaction 1) and a slow reaction (reaction 2). The second slope described only the slow reaction. The difference between the two slopes yielded the slope for reaction 1. Reaction 1 conformed to first-order kinetics for 8 min, at which time it began to terminate, leaving reaction 2 to proceed for many hours. Jardine and Sparks (1984a) attributed the two simultaneous first-order reactions at 283 and 298K in the Evesboro soil to exchange sites of varying K reactivity. Reaction 1 was ascribed to external surface sites of the organic and inorganic phases of the soil, which are readily accessible for cation exchange, whereas reaction 2 was attributed to less accessible sites of organic matter and interlayer sites of the 2:1 clay minerals that predominated in the <2 μm clay fraction.

In a recent study, Havlin *et al.* (1985) investigated K release from calcareous soils using a Ca-saturated resin. They found that first-order kinetics did not describe K release well. They noted that many workers who had found that the first-order equation described K dynamics well had investigated total reaction times that were < 2500 hr. In their study, they investigated times up to 7000 hr. They found that three first-order equations applied to segments of the total reaction time (1000, 1000 to 3000, 3000 to 7000 hr) were needed to describe K release kinetics in the soils they studied.

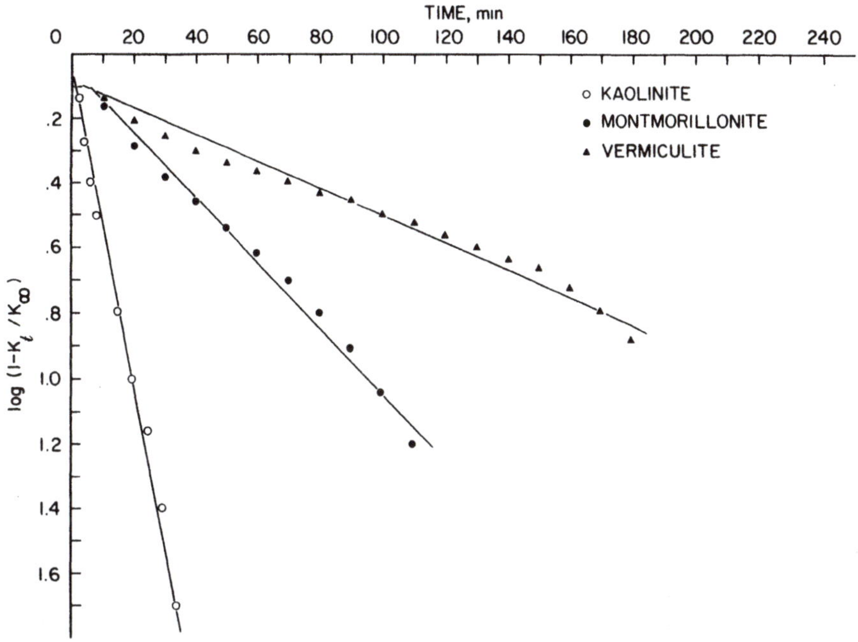


**Figure 2.** First-order plots of K adsorption on clay mineral systems. From Sparks and Jardine; © (1984) The Williams & Wilkins Co., Baltimore.

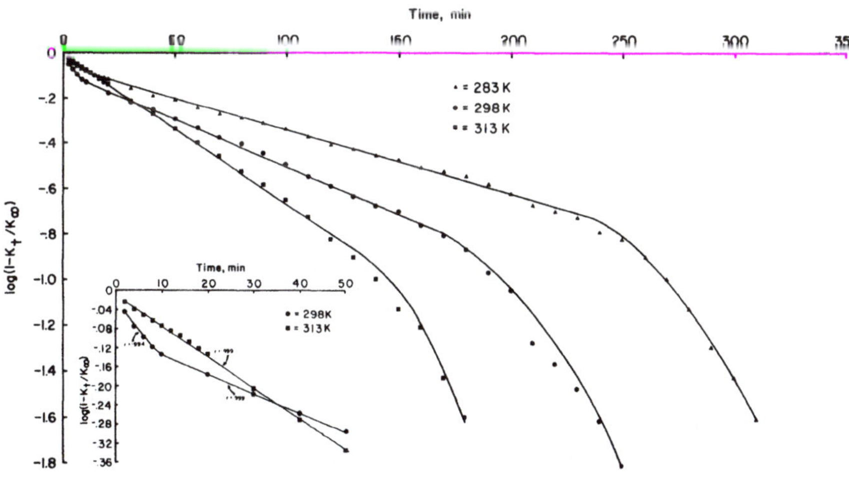


**Figure 3.** First-order kinetics of K adsorption at three temperatures on Evesboro soil with inset showing the initial 50 min of the first-order plots at three temperatures. From Jardine and Sparks; Reproduced from *Soil Science Society of America Journal,* Vol. 48, 1984, pp 39–45 by permission of the Soil Science Society of America.

## 2. Elovich Equation

The Elovich equation was first developed to describe the kinetics of chemisorption of gases on solid surfaces (Low, 1960). Recent theoretical studies with adsorption and desorption in oxide–aqueous solution systems showed that the applicability and method of fitting kinetic data to the Elovich equation require accurate data at short reaction times (Aharoni and Ungarish, 1976).

The Elovich equation has described the kinetics of heterogeneous isotopic exchange reactions well on goethite (Atkinson *et al.*, 1970), but not as well on gibbsite surfaces (Kyle *et al.*, 1975). More recently, a modified form of the Elovich equation has been applied successfully to kinetics of phosphate release and sorption (Chien and Clayton, 1980), dissolution of phosphate rocks in soils (Chien *et al.*, 1980), and dissolution of P to ethylenediamine tetraacetate (EDTA) (Onken and Matheson, 1982).

The Elovich equation applied to K release can be expressed (Havlin *et al.*, 1985) as:

$$dy/dt = ae^{-by} \qquad [1]$$

where $y$ is the amount of K released or adsorbed at time $t$, and $a$ and $b$ are constants (Low, 1960). The constant $a$ represents the initial rate because $dK/dt \rightarrow a$ as $y \rightarrow O$. Assuming $y = O$ at $t = O$, the integrated form of equation [1] is

$$y = 1/b \ln (1 + abt) \qquad [2]$$

or

$$y = 1/b \ln (t + t_o) - (1/b) \ln t_0 \qquad [3]$$

where $t_0 = 1/ab$. If $abt >>> 1$, then equation [2] can be simplified to:

$$y = (1/b) \ln (ab) + (1/b) \ln t \qquad [4]$$

Thus, a plot of $y$ vs. $\ln t$ for equation [4] should be linear with slope $1/b$ and intercept $1/b \ln (ab)$, where $b$ is the rate constant. If $abt$ is not $>>>$ 1, then a value of $t_0$ is chosen for equation [3] that results in a linear plot of $y$ vs. $\ln (t + t_0)$, with slope $1/b$ and intercept determined from $t_0 = 1/ab$.

Sparks and Jardine (1984) compared zero-order, first-order, parabolic diffusion, and the Elovich equations to describe K adsorption in clay minerals and soils. They found, using $r^2$ and standard error (SE) values, that the Elovich equation did not describe K adsorption well.

Havlin *et al.* (1985), investigating K release from calcareous soils, found that $abt \approx 1$ at $t < 1000$ hr, but $abt >>> 1$ at $t > 1000$ hr (see equation [4]). Therefore, the Elovich equation (assuming $abt >>> 1$; equation [4]) could not be used when $t < 1000$ hr. They concluded, therefore, that the Elovich equation should only be used when $abt >>> 1$ over the entire reaction period. When correct values of $t_0$ were chosen for the Elovich equation (equation [3]), then it described K release well. However, in its usual form, the Elovich equation described the data of Havlin *et al.* (1985) poorly.

### 3. Parabolic Diffusion Law

The parabolic diffusion equation for K dynamics can be expressed simply as:

$$K_t/K_\infty = Rt^{1/2} + \text{constant} \qquad [5]$$

where $K_t$ is the quantity of K adsorbed or released at time $t$, $K_\infty$ is the amount of K adsorbed or released at equilibrium, and $R$ is the overall diffusion coefficient and $t$ is time.

Numerous researchers have used the parabolic diffusion law to describe K dynamics in soils and clay minerals (Rausell-Colom *et al.*, 1965; Chute and Quirk, 1967; Quirk and Chute, 1968; Sivasubramaniam and Talibudeen, 1972; Sparks *et al.*, 1980c; Feigenbaum *et al.*, 1981; Jardine and Sparks, 1984a; Sparks and Jardine, 1984; Havlin *et al.*, 1985).

Sivasubramaniam and Talibudeen (1972) working on Al–K exchange on British soils, obtained parabolic plots that gave two distinct slopes, which the authors theorized could be indicative of two simultaneous diffusion-controlled reactions. The authors speculated that the rate-controlling step in the adsorption of $Al^{3+}$ and $K^+$ ions was the diffusion of the ions into the subsurface layers of the solid.

Sparks *et al.* (1980c) and Jardine and Sparks (1984a) noted a nonlinearity with the parabolic diffusion equation for the initial minutes of K desorption in soils. They attributed this deviation to film diffusion-controlled exchange in the early minutes of K exchange. However, Chute and Quirk (1967) ascribed the lack of conformity to the parabolic diffusion law of K release from micas in the early minutes of release to mass-action exchange.

Feigenbaum *et al.* (1981) successfully used the parabolic diffusion equation to describe K release from a number of K-bearing minerals (Figure 4) and calculated release rate coefficients ($k_2$) for muscovite of 0.44 $hr^{-1}$ for 5 to 20-μm mica particles and 0.38 $hr^{-1}$ for 20 to 50-μm particles.

### 4. Power-Function Equation

Havlin and Westfall (1985) and Havlin *et al.* (1985) used a power-function equation to describe K release from soils. The integrated form of the power-function equation can be expressed (Havlin and Westfall, 1985) as:

$$y = at^b \qquad [6]$$

The linear transformation is,

$$\ln Y = \ln a + b \ln t \qquad [7]$$

where $Y$ is the quantity of K released at time $t$, and $a$ and $b$ are constants. The $b$ value is a rate coefficient value.

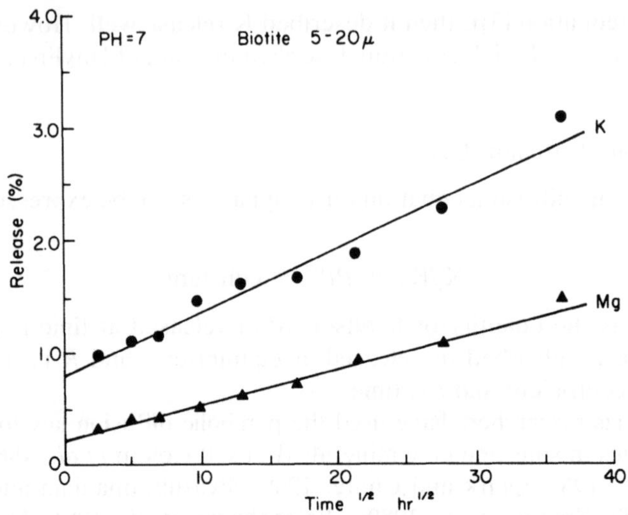


**Figure 4.** The rate of release of cations from biotite as a function of the square root of time under acidic and neutral conditions. From Feigenbaum *et al.;* Reproduced from *Soil Science Society of America Journal,* Vol. 45, 1981, pp 501–506 by permission of the Soil Science Society of America.

Havlin and Westfall (1985) found that the power-function equation described K release from soils well. The *a* and *b* values were highly correlated with nonexchangeable K release. As shown in Figure 5 they found that the rate coefficient (*b*) from equation [6] was highly correlated to K uptake and relative yield of alfalfa (*Medicago sativa* L.)

### B. Kinetic Models for Potassium

One approach to integrating and synthesizing a large body of knowledge on soil elemental transformations and transport is mathematical modeling and computer simulation. Excellent in-depth reviews on N modeling are given by Tanji (1982), on P modeling by Berkheiser *et al.* (1980) and Mansell and Selim (1981), on modeling waste water by Iskandar (1981), and on pesticide dynamics by Rao and Jessup (1982).

Unfortunately, few reports have appeared in the soil science literature on K modeling (Selim *et al.*, 1976; J. R. Jensen, 1984). There is a veritable need for dynamic models to predict fully the fate of K in soils with time. This would be particularly useful on the sandy soils of the world, where K leaching is often significant (Sparks *et al.*, 1980a). Perhaps Cooke and Gething (1978) best expressed the need for research on K modeling in soils when they addressed the International Potassium Congress at Bern, Switzerland in 1978:

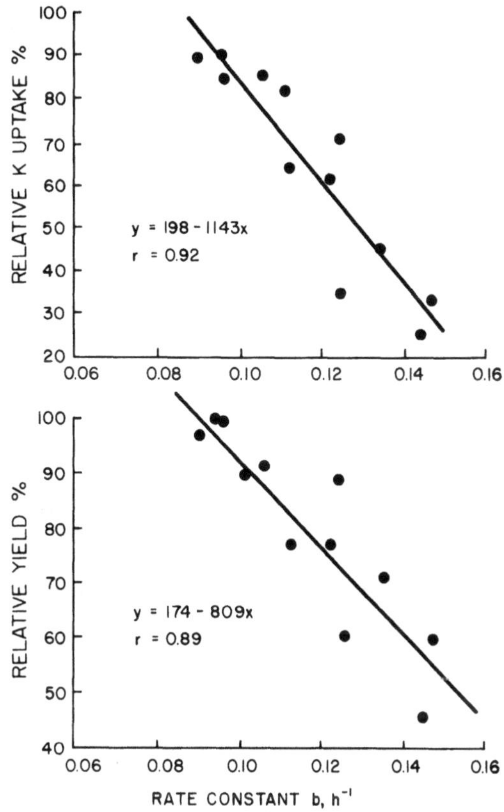


**Figure 5.** Relationship between relative K uptake and yield in greenhouse experiment and K release constant *b* as determined by Ca-resin extraction. From Havlin and Westfall; Reproduced from *Soil Science Society of America Journal*, Vol. 49, 1985, pp 366–370 by permission of the Soil Science Society of America.

It seems likely that further advances of the greatest significance for world farming will come from the development of dynamic mathematical models which can fully exploit our improved knowledge of the processes which govern the behavior of potassium in soils and in plant nutrition and which can forecast not only the effect of fertiliser treatment on the immediate crop but its effects on the future potential of the whole production system. The purpose of such models will be not only to ensure the best possible returns from using fertilisers but also to indicate how the nutrients in the whole crop-soil system may be conserved and used to the best advantages (pp. 361–362).

1. Model of Selim *et al.* (1976)

In 1976, Selim *et al.* proposed a mathematical model for K reactions and transport in soils. Kinetic reactions were assumed to govern the transformation between solution, exchangeable, nonexchangeable (secondary minerals), and primary mineral phases of K as illustrated in Figure 6.

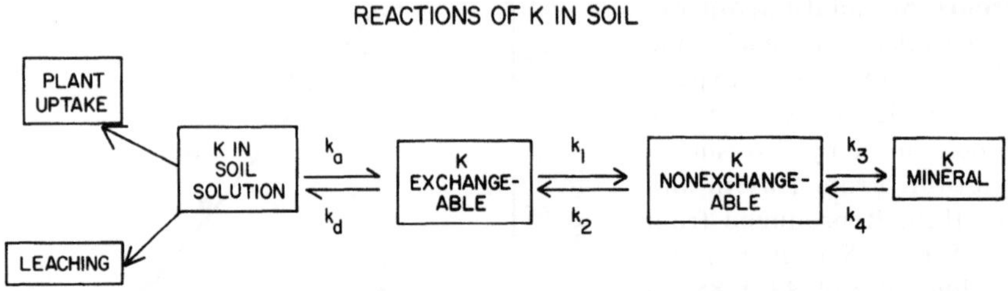


**Figure 6.** Kinetic reactions between phases of soil K where $k_a$ is the adsorption rate coefficient; $k_d$ is the desorption rate coefficient; $k_1$ is the nonexchangeable fixation rate coefficient; $k_2$ is the nonexchangeable release rate coefficient; $k_3$ is the immobilization rate coefficient; and $k_4$ is the mineralization rate coefficient. From Selim *et al.;* © (1976) The Williams & Wilkins Co., Baltimore.

Mathematical derivations of K transport and transformation processes may be formulated as follows. The following terms can be defined:

$C$ = concentration of K in solution phase
$S_1$ = amount of K in exchangeable phase per gram soil
$S_2$ = amount of K in nonexchangeable phase
$S_3$ = amount of K in primary mineral phase
$\rho$ = soil bulk density
$\Theta$ = soil water content
$v$ = pore water velocity
$D$ = dispersion coefficient
$t$ = time
$z$ = depth or distance below soil surface

The transport and transformation processes of K in the solution phase may be expressed (Selim *et al.*, 1976) as:

$$\frac{\partial c}{\partial t} = D\frac{\partial^2 c}{\partial Z^2} - \frac{v\partial C}{\partial z} - k_a C^N + (\rho/\partial)k_d S_1 \qquad [8]$$

The first two terms of the right side of equation [8] account for transport. These two terms are referred to as the dispersion and mass flow terms, respectively. The third and fourth terms represent the adsorption (forward) and desorption (backward) reactions between the exchangeable and solution phases. Selim *et al.* (1976) assumed that the backward reaction was a first-order, kinetic type, whereas the forward reaction was assumed to be $N^{th}$ order and less than unity.

The transformation of the exchangeable phase ($S_1$) may be written:

$$\frac{\partial S_1}{\partial t} = (\Theta/\rho)k_a C^N + k_2 S_2 - (k_1 k_d)S_1 \qquad [9]$$

Similarly, the transformations of the nonexchangeable phase ($S_2$) and of the mineral phase ($S_3$) may be expressed, respectively, as:

$$\frac{\partial S_2}{\partial t} = k_1S_1 + k_4S_3 - (k_2 + k_3)S_2 \qquad [10]$$

$$\frac{\partial S_3}{\partial t} = k_3S_2 - k_4S_3 \qquad [11]$$

In equations [9], [10], and [11], the reactions between exchangeable and nonexchangeable and those between nonexchangeable and primary mineral phases were assumed to be first-order kinetic reactions (Selim *et al.*, 1976).

The differential equations [8] to [11], which describe transport and reactions between K phases, may be regarded as a representation of the model shown in Figure 6. The initial and boundary conditions chosen (Selim *et al.*, 1976) were:

$$C = 0, \qquad t = 0, \qquad 0 < Z < L \qquad [12]$$

$$S_1 = S_2 = S_3 = 0, \qquad t = 0, \qquad 0 < Z < L \qquad [13]$$

$$vC - \frac{D\partial C}{\partial Z} = vC_0 \qquad t < t_1, z = 0 \qquad [14]$$

$$vC - \frac{D\partial C}{\partial Z} = 0 \qquad t > t_1, z = 0 \qquad [15]$$

$$\frac{\partial C}{\partial Z} = 0 \qquad t > 0, z = \mathrm{L} \qquad [16]$$

which describe K transport and transformation in a soil column of length $L$, initially ($t$ = O) devoid of K. A solution having soluble K concentration $C_0$ was applied at the soil surface for a length of time $t_1$ followed by a K free solution. A steady pore water velocity, $v$, throughout the soil, was assumed at all times.

The value of the parameter $N$ in equations [8] and [9] is different from unity, and so these differential equations are nonlinear and cannot be solved analytically. Therefore, equations [8] through [11] subject to conditions [12] through [16] were solved using numerical analysis techniques. Selim *et al.* (1976) used the explicit–implicit finite difference approximations as the method of solution. This was successfully used by Selim *et al.* (1975) for steady water flow conditions and by Selim *et al.* (1976) for transient water conditions. Finite difference approximations provide distributions of the various K phases ($C$, $S_1$, $S_2$, and $S_3$), at incremental distances $z$ in the soil and at discrete time steps $t$. A mass balance was maintained as a check on the numerical results.

## VII. Effect of Temperature on K Dynamics

That increasing the temperature frequently causes a marked increase in reaction rate has been known for many years. This relationship can best be described using the Arrhenius equation:

$$k = Ae^{-E_a/RT}$$

where $k$ is the rate coefficient, $A$ is the frequency factor, $E_a$ is the energy of activation, R is the universal gas constant, and T is absolute temperature. Thus, by plotting ln $k$ vs. $1/T$, one can calculate an $E_a$ value from the slope of the linear relationship.

Huang *et al.* (1968) investigated the effect of temperature on the rate of release of lattice K from four K-bearing minerals. A 10K rise in temperature during the reaction period resulted in a two- to threefold increase in the rate constant. For the temperature range used, the rate of K release from biotite was 13 to 16, 75 to 105, and 118 to 190 times higher than that from phlogopite, muscovite, and microline, respectively. Mortland (1958) observed a linear increase in the rate of release of K from vermiculite with increased temperature using 0.1$N$ NaCl as the leaching material.

Burns and Barber (1961) found that temperature increased the rate of release of nonexchangeable K to the exchangeable form for Indiana (USA) soils. The higher the temperature, the greater was the rate of release of nonexchangeable K.

Sparks and Jardine (1981) investigated the effect of temperature on the dynamics of K adsorption and desorption in a Matapeake soil (Table 8). The $k'_a$ and $k'_d$ values increased with increased temperature as predicted from Bronsted's reaction rate theory and from the Arrhenius equation.

# VIII. Dynamics Between Solution and Exchangeable Phases of Potassium

## A. Historical Perspective

J. Thomas Way (1850), the celebrated chemist who consulted with the Royal Agricultural Society, was the first researcher to propose that the rate of ion exchange on soils was instantaneous (Kelley, 1948; Thomas,

**Table 8.** Effect of temperature on apparent adsorption and desorption rate coefficients ($k'_a$ and $k'_d$, respectively)[a]

| Temperature (K) | $k_a'$ ($hr^{-1}$) | $k_d'$ ($hr^{-1}$) |
|---|---|---|
| 276 | 1.08 | 0.10 |
| 298 | 1.86 | 0.22 |
| 313 | 2.46 | 0.34 |
| | | |
| 276 | 0.72 | 0.08 |
| 298 | 1.35 | 0.19 |
| 313 | 1.74 | 0.27 |

[a]Adapted from Sparks and Jardine (1981).

1977). Gedroiz (1914) reported data for 100 g of Chernozem soil with 0.5 L of 1 mol NaCl $L^{-1}$ solution. After filtering, $Ca^{2+}$ was determined in the filtrate. These data showed that contact between the soil and solution for the brief period of 5 s plus the time required to filter the solution was enough time to reach equilibrium. However, the clay fraction of the Chernozem soil Gedroiz (1914) studied was largely montmorillonitic, which would be expected to exhibit rapid kinetic reactions (Malcom and Kennedy, 1969; Sparks and Huang, 1985).

As we shall see, ion-exchange kinetics are not always instantaneous, contrary to what the above investigators concluded. This is particularly true with K adsorption onto vermiculitic or micaceous soils and clay minerals.

**B. Rate of Exchange on Clay Minerals**

The reaction rate between soil solution and exchangeable phases of K is strongly dependent on the type of clay minerals present (Sivasubramaniam and Talibudeen, 1972; Sparks, 1980; Sparks *et al.*, 1980b,c; Sparks and Jardine, 1981, 1984; Sparks and Rechcigl, 1982; Jardine and Sparks, 1984a) and the method employed to measure the kinetics of K exchange (Sparks and Rechcigl, 1982; Sparks, 1985a,b; Sparks and Huang, 1985; Ogwada and Sparks, 1986a,b). Vermiculite, montmorillonite, kaolinite, and hydrous mica vary drastically in their ionic preferences, ion-binding affinities, and types of ion-exchange reactions. Such fundamental differences in these clay minerals account for the varying kinetics of exchange.

Kinetics of K exchange on kaolinite and montmorillonite are usually quite rapid (Malcom and Kennedy, 1969; Sparks and Jardine, 1984). An illustration of this is shown in Figure 7. In the case of kaolin clays, the tetrahedral sheets of adjacent clay sheets are held tightly by hydrogen bonds; therefore, only planar external surface (edge) sites are available for ionic exchange. With montmorillonite, the inner peripheral space is not held together by hydrogen bonds but instead is able to swell with adequate hydration and thus allow for rapid passage of ions into the interlayer. Malcolm and Kennedy (1969) found that the rate of Ba exchange on kaolinite and montmorillonite was rapid, with 75% of the total exchange occurring in 3 s.

Kinetics of K exchange on vermiculitic and micaceous minerals tend to be extremely slow. Both are 2:1 lattice clays with peripheral spaces that impede many ion-exchange reactions. Micaceous minerals typically have a more restrictive interlayer space than vermiculite because the area between layer silicates of the former is selective for certain types of cations (e.g., $K^+$, $Cs^+$). In 1963, Bolt *et al.* theorized the existence of three types of binding sites for K exchange on a "hydrous mica." The authors hypothesized that slow kinetics resulted with interlattice exchange sites, rapid kinetics with external planar sites, and intermediate kinetics with interlayer lattice edge sites.

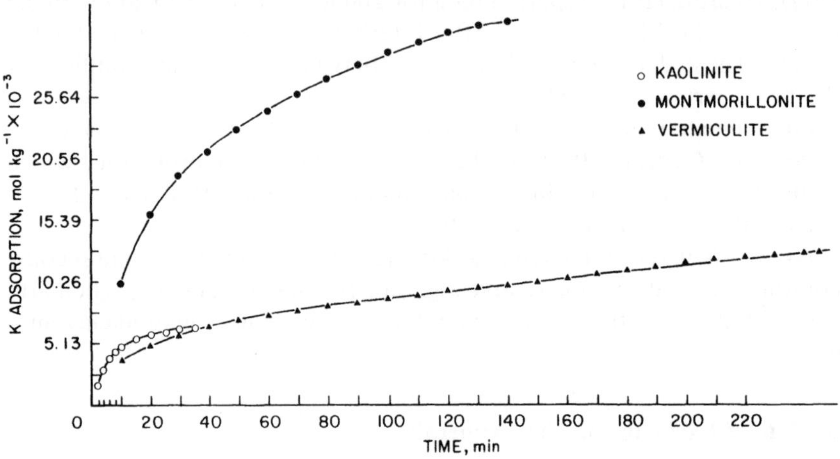


**Figure 7.** Potassium adsorption vs. time in clay systems. From Sparks and Jardine; © (1984) The Williams & Wilkins Co., Baltimore.

## C. Rate of Exchange on Soils

A number of papers have appeared in the soil science literature on the kinetics between solution and exchangeable forms of K in mixed soil systems (Talibudeen and Dey, 1968; Sivasubramaniam and Talibudeen, 1972; Goulding and Talibudeen, 1979; Goulding, 1980, 1981; Sparks *et al.*, 1980b,c; Sparks and Jardine, 1981, 1984; Sparks and Rechcigl, 1982; Jardine and Sparks, 1984a).

Sparks *et al.* (1980b) studied the kinetics of K adsorption in two Dothan soils (fine-loamy, siliceous, thermic Plinthic Paleudults) from Virginia. A slow rate of K exchange in these soils was ascribed to intraparticle transport and diffusion processes that reflected the relatively large quantities of vermiculitic material present in the soils. The $k_a$ values ranged from about 1 to 20 $hr^{-1}$, which suggested low rates of reaction as compared with values of 81 to 216 $hr^{-1}$ calculated for Florida soils (Selim *et al.*, 1976). This could be explained on the basis of the predominance of kaolinite in the Florida soils compared with vermiculitic minerals in the Virginia soils.

Sparks and Jardine (1981) investigated the kinetics of K adsorption and desorption in a Matapeake soil from Delaware. A diffusive process was noted and found to be slower for K desorption than for K adsorption. This would be expected because of the difficulty in desorbing K from partially collapsed interlayer positions of the vermiculitic clay minerals. Once K is adsorbed into this peripheral space, the coulombic attraction between $K^+$ ions and the silicate layers is greater than the hydration forces between the individual $K^+$ ions (Kittrick, 1966; Sawhney, 1966). Thus, a partial layer collapse occurs. The observation of slower desorption than adsorption suggests that K kinetics are nonsingular and that hysteresis may be occurring (Ardakani and McLaren, 1977; Rao and Davidson, 1978; Sparks *et al.*, 1980c). Lower rates for K desorption have also been noted by others (Talibudeen and Dey, 1968; Feigenbaum and Levy, 1977).

Jardine and Sparks (1984a) found that the kinetics of K adsorption and desorption in an Evesboro soil were explained by two simultaneous first-order reactions at 283 and 298K, which were ascribed to sites of varying K reactivity (see Figure 3). Reaction 1 was ascribed to external surface sites of the organic and inorganic phases of the soil that are readily accessible for cation exchange, whereas reaction 2 was attributable to interlayer sites of the 2 : 1 clay minerals that predominated in the <2-μm clay fraction.

### D. Use of Electroultrafiltration for Determining the Dynamics of K Exchange

A discussion of general kinetic methodologies for determining the dynamics of K reactions in soils and clay minerals has been given earlier (Section IV). These were subdivided into batch and flow techniques. Another method, which is a combination of both flow and batch techniques and which has been useful in studying K dynamics and K availability in soils, is electroultrafiltration (EUF) (Nemeth, 1971a; Nemeth, 1979; Mengel, 1985).

The EUF method is essentially a combination of electrodialysis and ultrafiltration. Originally it was used for the purification of colloidal suspensions (Grimme, 1979). Later this procedure was adopted and modified by Kottgen and Diehl (1929) for soil testing purposes. Their idea was to extract the labile fraction of plant nutrients without having to treat the soil with chemicals. The EUF method avoids some of the shortcomings of electrodialysis, such as the accumulation of reaction products in the electrode compartments, the buildup of charge on the membranes separating the compartments, and the effect of electroosmosis (Nemeth, 1972; Grimme, 1979).

The EUF method was again modified by Nemeth (1971a, 1972), who, instead of one constant field strength, used three different field strengths on one sample. This modified procedure has some practical advantages, especially if information about the nutrient status of soils with respect to availability is required.

The EUF method not only allows the determination of intensity, quantity, and buffering parameters (Nemeth, 1976; Grimme, 1980), but it also allows an assessment of the change of mobile soil K with time (Figure 8).

The first two fractions (5 to 10 min) are a measure of K intensity, and K extracted within this time is correlated with the K concentration in the soil solution, whereas the sum of all fractions is a quantity parameter.

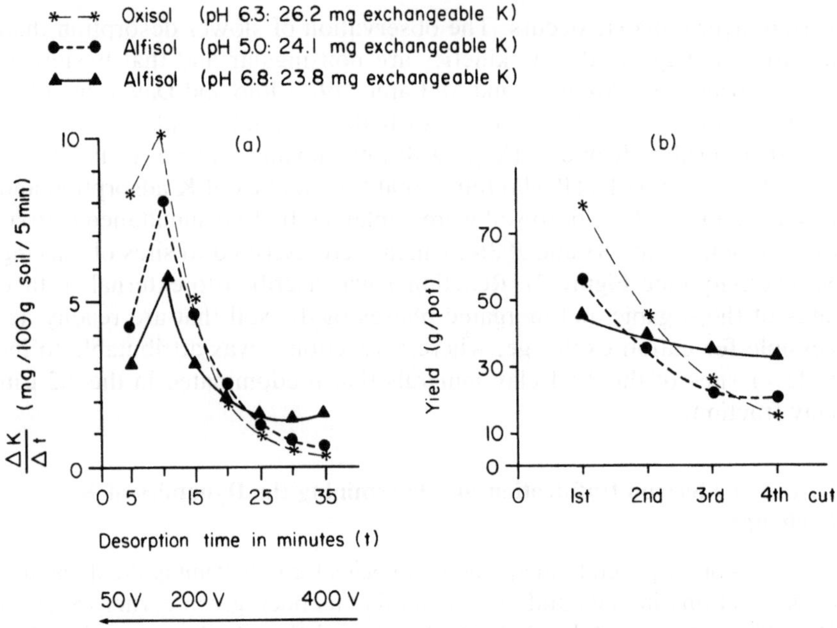


**Figure 8.** Comparison of EUF desorption characteristics of three different soils (a) and the decrease of ryegrass yield with time (b). The soils contain similar amounts of exchangeable K but differ in K selectivity and buffering power. The yield curves are parallel to the desorption curves (Grimme, 1980).

The slopes of the desorption curves indicate the rate with which the available quantity will be exhausted. Therefore, the ratio of the seventh and second fraction is a very convenient measure of the K buffer power of the soil. Using EUF analysis, one can thus obtain a good estimate of exchangeable K. Grimme (1980) found that there was no evidence of a significant release of nonexchangeable K during EUF analysis. Additionally, in his study, there was no destruction of K-bearing minerals during K desorption.

# IX. Kinetics of Nonexchangeable and Mineral Potassium Release

## A. Dynamics of K Release from Micas

### 1. Mechanisms of K Release

The release of K from micas proceeds by two processes: (1) the transformation of K-bearing micas to expansible 2 : 1 layer silicates by exchanging the K with hydrated cations and (2) the dissolution of the micas followed by the formation of weathering products (Sparks and Huang, 1985). The relative importance of these two mechanisms depends on the stability of micas and the nature of soil environments.

The release of K through the transformation of micas to expansible 2 : 1 layer silicates has been viewed to take place by (1) edge weathering (Mortland, 1958; Rausell-Colom *et al.*, 1965; Scott and Smith, 1967) and (2) layer weathering (preferential weathering planes, Jackson *et al.*, 1952). Three "mica–vermiculite," particles, each with about 50% expansible layers, are shown in Figure 9. In models a and b, the expanded interlayers are not continuous but have internal terminations forming "wedge zones" (Rich and Black, 1964; Rich, 1972). These "wedge zones" are important in K selectivity because they are selective for such ions as $[NH_4]^+$, $[H_3O]^+$, and $K^+$ and screen out such large ions as $Ca^{2+}$ and $Mg^{2+}$ (Sparks and Huang, 1985). As more interlayers are partly opened, wedge zones increase and then decrease as interlayers are open all the way through the particle. In model c interlayer spaces are either expanded or nonexpanded uniformly through each layer. All models apparently apply. Layer weathering appears to be most common with small mica particles, whereas edge weathering is most common in larger particles (Scott, 1968; Huff, 1972; Norrish, 1973).

The mechanisms of the release of K from micas by the dissolution of

**Figure 9.** Diagram of a weathered mica particle containing about 50% expanded (vermiculite) layers; (a), "frayed edge" and mica core; (b), alternate layers open halfway through interlayer; and (c), regularly interstratified mica–vermiculite (no wedge zones). From Rich (1972).

the minerals (Huang *et al.*, 1968; Kittrick, 1973) are more complex than simple transformations of micas to 1 : 1 expansible layer silicates. However, this type of reaction occurs in soil environments.

A number of factors govern the release of K from micas by both cation-exchange reactions and dissolution processes. These include tetrahedral rotation and cell dimensions, degree of tetrahedral tilting, hydroxyl orientation, chemical composition, particle size, structural imperfections, degree of K depletion, layer charge alterations and associated reactions, hydronium ions, biological activity, inorganic cations, weathering, and drying. These factors are not discussed in detail here but are described fully by Sparks and Huang (1985).

### 2. Specific Studies

Much work has been performed on the release of interlayer K from trioctahedral micas (Walker, 1950; Mortland *et al.*, 1956; Arnold, 1958; Mortland, 1958; Basset, 1960; Reed and Scott, 1962; Scott and Reed, 1962; Quirk and Chute, 1968; Wells and Norrish, 1968; Malquori *et al.*, 1975; Feigenbaum *et al.*, 1981). It is convenient to work with trioctahedral micas because their interlayer K is more easily removed than the interlayer K in dioctahedral micas (Reed and Scott, 1962). The trioctahedral micas are also much more subject to acid dissolution than their dioctahedral counterparts (Wells and Norrish, 1968). The more unstable trioctahedral mica–ferruginous biotite can be completely broken down by H-saturated resin in about 10 days, releasing all of its K (Arnold, 1958). This explains the much higher rate constant for K release from these micas than has been observed for dioctahedral micas. Because of their instability, trioctahedral micas generally occur only in soils where little weathering has taken place, and so such soils contain large quantities of nonexchangeable K (Milford and Jackson, 1960).

Feigenbaum *et al.* (1981) investigated the rate of release of K and structural cations from biotite, phlogopite, and muscovite in two particle-size ranges in dilute electrolyte solutions (0.001 mol $L^{-1}$) and at pH 3.0 and 7.0. The rate of K release from phlogopite and biotite was similar to the rate of release of other structural cations under acidic conditions and significantly higher under neutral conditions (Table 9). The authors felt these findings indicated that structural decomposition of phlogopite and biotite was dominant in acidic conditions, and the rate of interdiffusion increased in neutral conditions. The rate of K release from muscovite was about 5 and 15% that from biotite and phlogopite, respectively (Table 9). The rate of K release from muscovite was higher than the rate of $Al^{3+}$ release. Feigenbaum *et al.* (1981) concluded that muscovite was the most stable of the three micas and that the decomposition mechanism for K release in muscovite was less important.

**Table 9.** Rate of release of K and structural cations from the mica minerals[a]

| Mineral | Particle size (mm) | Parameter[b] | Acidic conditions[c], pH = 3.0 | | | | | Neutral conditions[c], pH = 7.0 | | |
|---|---|---|---|---|---|---|---|---|---|---|
| | | | K | Mg | Al | Fe | Mn | K | Mg | Mn |
| Phlogopite | 5–20 | Slope | 2.80a | 2.16a | 2.15a | 2.66a | 2.80a | 0.027a | 0.013b | |
| | 20–50 | Slope | 1.82a | 1.70a | 1.34a | 1.86a | 2.00a | 0.021a | 0.007b | |
| Biotite | 5–20 | Slope | 8.02a | 9.05a | — | 6.53b | 8.20a | 0.056a | 0.033b | 0.067a |
| | 20–50 | Slope | 7.09a | 8.12a | — | 5.68a | 7.66a | 0.033a | 0.010b | 0.028a |
| Muscovite | 5–20 | Slope | 0.44b | 0.97a | 0.12b | 0.42b | — | | | |
| | 20–50 | Slope | 0.38b | 0.83a | 0.19b | 0.20b | — | | | |

[a]Permission granted from Soil Science Society of America, Madison, Wisconsin. Feigenbaum *et al.* (1981).

[b]Slopes of the straight lines relating the fraction of released cation (%) to the square root of time, $h^{1/2}$ (parabolic diffusion equation).

[c]Different letters designate significantly different values at the 0.05 level of probability.

### B. Dynamics of Potassium Release from Feldspars

The predominance of the feldspars among minerals of the lithosphere makes their disappearance by weathering quantitatively the most important chemical reaction since the solidification of the earth's crust (Marshall, 1964; Sparks and Huang, 1985). The kinetics and particularly the mechanisms of K release from feldspars are not well understood (Rasmussen, 1972; Berner, 1983; Sparks and Huang, 1985). However, a number of studies on these topics have been conducted and are elaborated on below.

#### 1. Mechanisms of Potassium Release

One of the first workers to apply true chemical kinetics to weathering products was Wollast (1967), who described the results of experiments in which ground K-feldspar was placed in buffered aqueous solutions and the release of silica with time was measured. Wollast (1967) found that the rate of addition of silica to solution decreased with time and he explained this decrease in terms of the formation of a protective layer of aluminous precipitate on the surface of the feldspar. The rate of dissolution was assumed to be controlled by the rate at which silica and cations could diffuse through the surface layer (Berner, 1983).

Subsequent workers (Helgeson, 1971; Busenberg and Clemency, 1976) have expanded on the Wollast model and have suggested that the release of Si and cations from alkali feldspars can be described in terms of parabolic kinetics, where the concentration in solution builds up in direct proportion to the square root of time. The protective surface layer is presumed to be either a precipitate of aluminum hydroxide and/or clay minerals, or "feldspar" in which the cations have been replaced by hydrogen or $[H_3O]^+$ ions.

The concepts of parabolic kinetics and a protective surface layer formed on alkali feldspar during weathering have been tested by subsequent work (Petrovic, 1976; Petrovic *et al.*, 1976; Holdren and Berner, 1979; Berner and Holdren, 1979). In the studies by Petrovic *et al.* and by Holdren and Berner, ground sanidine albite was subjected to leaching by aqueous solutions in the laboratory under conditions essentially identical to those employed by Wollast. Parabolic release of silica was found but, unexpectedly, the surfaces of the reacted feldspar grains were not found altered in any way. The Wollast model predicts an altered surface of several hundred Angstroms thickness, but using x-ray photoelectron spectroscopy (XPS), Berner (1983) found that no alteration product any thicker than about 5 to 15 Å could be present on the feldspar surface. Additionally, Berner (1983) found that the parabolic kinetics could not be reproduced if the ground feldspar grains used for experimentation were first treated with HF to remove ultrafine and strained particles produced by grinding. Berner (1983) obtained only linear kinetics, or a constant rate of release of silica to solution with time, with HF-treated material. Berner (1983) contends that the "parabolic" kinetics obtained by other workers is the result of varying rates of dissolution of particles produced during grinding.

Berner and Holdren (1979), by examining feldspar grains taken from soils, essentially verified their hypotheses that feldspar dissolved according to linear kinetics and did not produce an altered surface layer. Using scanning electron microscopy (SEM) analyses, they found clay adhering to the surfaces of the feldspar grains, but the authors speculated that this clay could not have been protective in the sense envisioned by Wollast (1967). Berner and Holdren (1979) used XPS to examine the surfaces of ultrasonically cleaned feldspar grains and found that the surface composition in the outermost few tens of Angstroms was essentially identical to the grain interiors. They found no cation depletion or Al enrichment as would be expected for an altered surface layer.

If feldspar dissolution during weathering is not controlled by diffusion through a protective surface layer, then how does it dissolve? Recent work (Dibble and Tiller, 1981; Berner, 1983) would indicate that the limiting step in dissolution is not diffusion but chemical reaction at the feldspar–water interface. A number of studies examining feldspar grains using SEM show the growth and development of characteristic, crystallographically controlled etch pits on the surface, which indicates dissolution rate control by surface chemical reaction (Wilson, 1975, Berner and Holdren, 1979; Mirecki *et al.*, 1985). An example is shown in Figure 10. The etch pits indicate selective attack of the feldspar surface by soil acids at points of excess energy, such as dissolution outcrops. General attack of the surface with consequent rounding, as predicted for diffusion-controlled dissolution, is not found (Berner, 1983).

## 2. Specific Studies

The chemical dynamics of dissolution of K from K-feldspar by a 1 mol $L^{-1}$ $HNO_3$ solution was studied by Huang *et al.* (1968). For the temperature range studied, the rates of K release from muscovite, phlogopite, and biotite were about 2, 9 to 12, and 118 to 190 times, respectively, greater than the rate of K release from microcline (Table 10). The release of K from a tektosilicate is more difficult than from the interlayers of the micas. The Arrhenius heat of activation of K from microcline was 96.24 kJ $mol^{-1}$, which was greater than for the micas. The $E_a$ is interpreted as the energy level the structural K must acquire to be able to react. As the $E_a$ increases the rate constant decreases, because the mineral with a higher heat of activation for the release of structural K would release less K per unit time. The data of Huang *et al.* (1968) indicate that the release of K from K-feldspars is not controlled by film diffusion, because the $E_a$ values for diffusion-controlled reactions range from 17 to 34 kJ $mol^{-1}$ (Glasstone *et al.*, 1941). The magnitude of the values obtained by Huang *et al.* (1968) suggests a chemical reaction mechanism for K release from feldspars, as Berner (1983) has found.

**Figure 10.** Etch pits on feldspar grains from the fine sand fraction of a Delaware soil. From Mirecki *et al.* (1985).

Very recently, Song and Huang (1983) found that the sequence of K release from K-bearing minerals by oxalic and citric acids is biotite > microcline ≈ orthoclase > muscovite (Table 11). This differs from the sequence that occurs through the cation reaction using sodium tetraphenylboron (NaTPB).

**Table 10.** Apparent rate constants and Arrhenius heats of activation for the release of structural K from K minerals[a]

| Mineral | Rate Constant ($hr^{-1}$) at: 301K | 311K | Arrhenius heat of activation ($kJ\ mol^{-1}$) |
|---|---|---|---|
| Biotite | $1.46 \times 10^{-2}$ | $3.09 \times 10^{-2}$ | 58.66 |
| Phlogopite | $9.01 \times 10^{-4}$ | $2.44 \times 10^{-3}$ | 77.81 |
| Muscovite | $1.39 \times 10^{-4}$ | $4.15 \times 10^{-4}$ | 85.52 |
| Microcline | $7.67 \times 10^{-5}$ | $2.63 \times 10^{-4}$ | 96.24 |

[a]Huang *et al.* (1968). Permission granted from Author.

**Table 11.** Potassium released from K-bearing minerals by various solutions at the end of a 10-day reaction period[a]

| Mineral | Organic acid[b] (g of K released $kg^{-1}$ of structural K) | | |
|---|---|---|---|
| | Oxalic | Citric | NaTPB[c] |
| Biotite | 44 | 11 | 1000 |
| Muscovite | 1.1 | 0.9 | 83 |
| Microcline | 2.3 | 1.4 | 1.6 |
| Orthoclase | 2.2 | 1.2 | 1.3 |

[a]Song and Huang (1983).

[b]Concentration of organic acid solutions was 0.01 mol $L^{-1}$.

[c]1 mol $L^{-1}$ of NaCl + 0.2 mol $L^{-1}$ of NaTPB + 0.01 mol $L^{-1}$ of EDTA.

### C. Dynamics of Nonexchangeable Potassium Release from Soils

Few studies have been conducted on the kinetics of nonexchangeable K release from whole soils (Talibudeen and Weir, 1972; Talibudeen *et al.*, 1978; Martin and Sparks, 1983; Havlin and Westfall, 1985; Havlin *et al.*, 1985; Sadusky and Sparks, 1985).

Martin and Sparks (1983) investigated the kinetics of nonexchangeable K release from two coastal plain soils from Delaware using an H-resin (Figure 11).

Equilibrium in K release was attained in about 40 days for the two soils. Nonexchangeable K release rate coefficients ($k_2$) were calculated for all depths of the two soils investigated by Martin and Sparks (1983). They ranged from 1.2 to 2.2 $\times$ $10^{-3}$ $hr^{-1}$ in the Kalmia soil (a fine-loamy over sandy or sandy-skeletal, siliceous, thermic Typic Hapludult) and from 1.5 to 2.9 $\times$ $10^{-3}$ $hr^{-1}$ in the Kennansville soil (loamy, siliceous, Thermic Arenic Hapludult).

Recently, Sadusky and Sparks (1985) investigated the rate of nonexchangeable K release from sandy Atlantic Coastal Plain soils high in K-feldspars using a H resin and oxalic acid. The kinetics of K release were studied on Ca-saturated samples of the whole soil and of the sand fractions. Considerable K was released from the sand fractions using H resin (Table 12), whereas less K was released using oxalic acid. The data of Sadusky and Sparks (1985) point out the importance of even the sand fractions of some soils in supplying K to plants. Further attention must be given to the K-bearing minerals in sand and silt fractions of soils as they affect K dynamics.

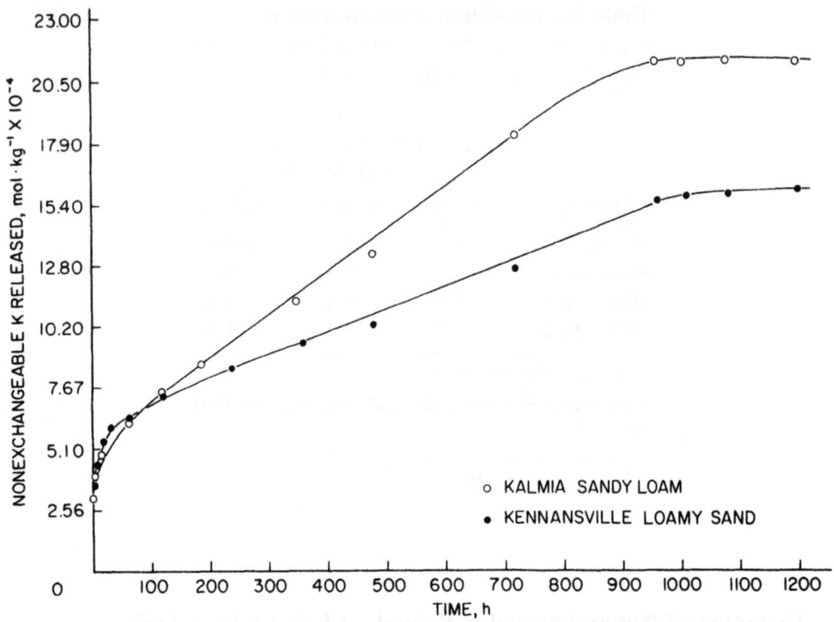


**Figure 11.** Amount of nonexchangeable K released vs. time in the 0.45- to 0.60-m depth of Kalmia and Kennansville soils. From Martin and Sparks; Reproduced from *Soil Science Society of America Journal,* Vol 47, 1983, pp 883–887 by permission of the Soil Science Society of America.

## D. Methods for Determining the Dynamics of Nonexchangeable Potassium

A number of methods and extractants have been used to study the rate of nonexchangeable K release from soils and clay minerals. These have included extractions with salts, acids, ion-exchange resins, NaTPB, electric current, and organic acids (Martin and Sparks, 1983; Sparks and Huang, 1985).

### 1. Acids and Salts

One of the quickest and easiest ways of determining the dynamics of nonexchangeable K release from soils is with boiling $HNO_3$. Most workers boil the soil in 1 mol $HNO_3$ $^{-1}$ per liter for 10 min over a flame, transfer the slurry to a filter, leach the soil with dilute $HNO_3$, and then determine the K content of the extract. This method has been described by Pratt (1965).

Huang *et al.* (1968) did not boil the slurry but allowed it to stand at 301 and 311K for various periods of time. MacLean (1961) used overnight soaking in 0.1 mol $HNO_3$ per liter and repeated boiling, extracting more K than the regular procedure would. One of the problems with boiling

**Table 12.** Potassium release from sand fractions of a Delaware soil using a H-resin and oxalic acid[a]

| | K Released from Sand Fractions mg kg$^{-1}$ [b] | | | | | |
|---|---|---|---|---|---|---|
| | Coarse (0.50–1.00 mm) | | Medium (0.25–0.50 mm) | | Fine (0.10–0.25 mm) | |
| Horizon | H-resin | Oxalic acid | H-resin | Oxalic acid | H-resin | Oxalic acid |
| Ap | 53.5 | 0.50 | 65.0 | 0.72 | 71.5 | 1.05 |
| E | 53.5 | 0.55 | 58.0 | 0.72 | 87.5 | 0.99 |
| Bt2 | 76.0 | 1.06 | 69.5 | 1.05 | 99.5 | 1.10 |
| C | 87.5 | 1.27 | 88.5 | 1.27 | 99.5 | 1.76 |

[a]Sadusky and Sparks (1985).

[b]These values represent amounts of K released at 30 days.

only 10 min over flame (Pratt, 1965) is that it is difficult to be precise about the correct boiling time, the time it takes for boiling to occur, and the vigor of boiling. To avoid this problem, Pratt and Morse (1954), Pratt (1965), and Conyers and McLean (1969) have used a 386K oil bath for 25 min, including heating time. This releases the same amount of K as does a flame but is more precise and easier when large numbers of samples must be handled. The main problem with boiling $HNO_3$ and other strong acids for soils is their potential for dissolution of mineral forms of K (Barber and Mathews, 1962; Martin and Sparks, 1983).

Other researchers have used continuous leaching with dilute acids (Mortland, 1958), such as 0.01 mol HCl per liter (Garman, 1957; MacLean, 1961), or with electrolyte solutions, such as 0.1 mol NaCl per liter (Mortland and Ellis, 1959), repeated extractions with 3, 0.3, and 0.03 mol NaCl per liter (Quirk and Chute, 1968), Sr salts (Wells and Norrish, 1968), hot $MgCl_2$ (White, 1950), and sodium cobaltinitrate (White, 1950).

## 2. Cation-Exchange Resins

The use of cation-exchange resins to simulate the uptake of nonexchangeable K by plants was suggested by Wiklander (1949). Hydrogen-saturated resins have been used for this purpose by Pratt (1951), Schmitz and Pratt (1953), Salomon and Smith (1957), Arnold (1958), MacLean (1961), Stahlberg (1959), Scott *et al.* (1960), Barber and Mathews (1962), Haagsma and Miller (1963), Feigenbaum *et al.* (1981), and Martin and Sparks (1983). These resins have high cation-exchange capacities, far exceeding those of soils. When saturated with an appropriate cation and mixed with soil and with a dilute solution of some sort, they can adsorb and hold all of the K released from the soil.

Calcium- and Na-saturated resins have been tried and found unsatisfactory by Arnold (1958), Stahlberg (1959), Haagsma and Miller (1963), and Feigenbaum *et al.* (1981) when used with any soil minerals more stable than trioctahedral micas. However, Talibudeen *et al.* (1978) argued that H-saturated resin might be destructive to soil minerals and consequently used Ca-saturated resin. After ≈100 hours of equilibration, the resin could not absorb further K and K release stopped. Talibudeen *et al.* (1978) ameliorated this problem by separating the resin and the soil before further K release stopped and then adding a new charge of resin. The separation process, however, seemed to have caused some exfoliation of clay minerals during dispersion in deionized water.

Recently, Havlin and Westfall (1985) determined K release kinetics and plant response in a number of calcareous soils. They found that nonexchangeable K release to alfalfa (*Medicago sativa* L.) and to Ca-resin were highly correlated (Table 13).

The question of the rate of $[H_3O]^+$ ions in nonexchangeable K release and soil mineral weathering is surely important. Arnold (1958) found muscovite and hydrous mica to be comparatively resistant to H-resin attack. The replacement of interlayer K has been shown to be unaffected by pH changes in the range of 4.6 to 9.2 (S.J. Smith and Scott, 1966), 4 to 8 (Chute and Quirk, 1968), 3 to 6.8 (Hossner, 1966), and 3 and above (Wells and Norrish, 1968). Haagsma (1961) found that little acid decomposition of soil minerals took place above pH 2.5 in a soil–resin mixture. Wells and Norrish (1968) showed that the $[H_3O]^+$ ion behaved like a metal cation with regard to K replacement. Norrish (1973) further states that in very weak acid concentrations ($10^{-3}$ mol $L^{-1}$), the $[H_3O]^+$ ion behaves as any other cation in replacing interlayer K and that only with higher concentrations of acid is the octahedral sheet attacked and its structure destroyed. Martin and Sparks (1983) found that H-saturated resin did not appear to cause release of significant mineral K from two Atlantic Coastal Plain

**Table 13.** Correlation coefficients (*r*) of nonexchangeable K release with plant response and soil test K levels[a]

| | Nonexchangeable K release | |
|---|---|---|
| | Greenhouse (K balance)[b] | Laboratory (Ca-resin)[b] |
| Cumulative yield | 0.96 | 0.87 |
| Cumulative K uptake | 0.99 | 0.94 |
| Initial $NH_4OAc$-K | 0.94 | 0.92 |
| Steady state $NH_4OAc$-K | 0.96 | 0.84 |

[a]From Havlin and Westfall (1985).
[b]Significant at 0.01 probability level.

soils. Huang *et al.* (1968) justify mild acid treatment for measuring K release on the basis of Keller's (1955) statement that the rhizosphere ionic atmosphere is dominated by $[H_3O]^+$. The generally accepted notion that the rhizosphere pH is lower than that in the bulk soil has been seriously challenged by Nye (1981). He provides evidence for the rhizosphere pH being 1 to 2 units higher than the bulk soil. This issue remains unresolved.

In order for electrolyte solutions and cation-exchange resins to be effective, the K concentration in the solution phase must be kept very low, or K release is inhibited (Bassett, 1959; Rausell-Colom *et al.*, 1965; S.J. Smith and Scott, 1966; Norrish, 1967; Scott and Smith, 1967; Wells and Norrish, 1968; Feigenbaum *et al.*, 1981). The critical concentration above which release is inhibited has been reported as 4 $\mu g\ ml^{-1}$ (S.J. Smith and Scott, 1966) in general, 2.3 to 16.8 $\mu g\ ml^{-1}$ for trioctahedral micas in dilute solution, and as low as $<0.1\ \mu g\ ml^{-1}$ for muscovite and illite (Rausell-Colom *et al.*, 1965). Maintenance of a low enough concentration of K can be accomplished with continuous flow of extracting or exchanging solution (Rausell-Colom *et al.*, 1965; Wells and Norrish, 1968), with cation-exchange resins (Arnold, 1958; Feigenbaum *et al.*, 1981), or with NaTPB (Scott *et al.*, 1960).

### 3. Sodium Tetraphenylboron

The NaTPB method was developed by Scott *et al.* (1960) and has been used also by Scott and Reed (1962), Reed and Scott (1962), Scott (1968), Conyers and McLean (1969), and Ross (1961). The $[BPh_4]$- anion combines with released K in solution and precipitates, while the $Na^+$ ion acts as an exchanger for interlayer K.

### 4. Electric Currents

Another technique that has been used for nonexchangeable K analysis is electrodialysis. It has been used by Peech and Bradfield (1934), Gilligan (1936), Ayres *et al.* (1947), and Reitemeier (1951). A soil slurry is subjected to a current, usually 110 V, for various lengths of time, causing various forms of K to be released into solution. Barber and Mathews (1962) warned that electrodialysis might break down K minerals excessively.

### 5. Organic Acids

Biological activity promotes K release from micas (Mortland *et al.*, 1956; Boyle *et al.*, 1967; Weed *et al.*, 1969; Sawhney and Voight, 1969; Sparks and Huang, 1985). The organisms deplete the K of the soil solution, and their action may be linked to that of tetraphenylboron (TPB) in artificial weathering of micas.

Recently, Song and Huang (1983) have studied the influence of oxalic and citric acids on the dynamics of K release from K-bearing minerals. They found that the sequence of K release was biotite > microcline ≈

orthoclase > muscovite (Table 11). This differed from the sequence that occurs through cation reaction using NaTPB. Sadusky and Sparks (1985) used oxalic acid and a H-resin to study K release from Delaware soils and from the sand fractions of these soils. Much higher quantities of K were released with the H-resin than with oxalic acid.

## X. Use of Kinetics for Determining Thermodynamics of Potassium Exchange in Soils

It has been known for many years that a direct relationship exists between chemical equilibria and completely reversible reactions (e.g., see Glasstone *et al.*, 1941; Laidler, 1965). Keay and Wild (1961) used this approach to calculate the enthalpy and entropy of Na–Mg exchange on vermiculite from measurements of kinetics of Mg–Na and Na–Mg exchange. Sparks and Jardine (1981) first applied the relationship to K-exchange studies, and they calculated apparent thermodynamic parameters of K–Ca exchange on a soil from measurements of the kinetics of this reaction. A miscible displacement technique (Sparks *et al.*, 1980c) was used to calculate the kinetic parameters. The magnitude of the apparent thermodynamic parameters determined by Sparks and Jardine (1981) compared well to pseudothermodynamic parameters calculated using Eyring's reaction rate theory (Table 14); however, the authors clearly referred to the equilibrium constant they calculated ($k'_a/k'_d$) as an apparent one ($K'_{eq}$). Moreover, the authors did not also calculate thermodynamic parameters from classical exchange isotherm data and compare the results using the two theories.

Ogwada and Sparks (1986a) determined the theoretical and practical implications of using a kinetic approach to calculate thermodynamic parameters for K–Ca exchange in soils. This was accomplished by comparing both in magnitude and trend the parameters using kinetic and equilibrium theories, and their study is elaborated upon later (Section XB).

### A. Theoretical Considerations

#### 1. Equilibrium Approach

If one employs the classical equilibrium theory of Argersinger *et al.* (1950) to the following exchange reaction:

$$\mathrm{Ca-(Soil)_2} + 2\,\mathrm{K^+} \leftrightharpoons 2\,\mathrm{K-Soil} + \mathrm{Ca^{2+}} \qquad [18]$$

one finds that the thermodynamic equilibrium constant ($K_{eq}$) is given by:

$$\mathrm{K_{eq}} = \frac{a^2_{\mathrm{K}(ad)}\, a_{\mathrm{Ca}}}{a_{\mathrm{Ca}(ad)}\, a^2_{\mathrm{K}}} \qquad [19]$$

**Table 14.** Comparison of $\Delta G°$ and $\Delta H°$ values for potassium exchange in Matapeake soil using a kinetic approach and Eyring's reaction rate theory[a]

| Temperature | $\Delta G°$ as established from a kinetic approach | $\Delta G°$ as established from Eyring's reaction rate theory | $\Delta H°$ as established from a kinetic approach | $\Delta H°$ as established from Eyring's reaction rate theory |
|---|---|---|---|---|
| 276 | −5422 | −5447 | | −7081 |
| 298 | −5338 | −5405 | −7081 | −7081 |
| 311 | −5187 | −5196 | | −7081 |
| 276 | −4936 | −4986 | | −5615 |
| 298 | −4840 | −4819 | −5615 | −5615 |
| 311 | −4852 | −5860 | | −5615 |

[a]from Sparks and and Jardine (1981). Permission granted from Soil Science Society of America, Madison, Wisconsin.

where $a$ ($_{ad}$) is the activity of the cation on the exchanger phase and $a$ is the activity of the cation in the solution phase.

Treating the exchanger as a nonideal solid solution and assuming the activity of adsorbed water is negligible,

$$a^2_{K(ad)} = f^2_K \overline{N}^2_K \quad \text{and} \quad a_{Ca(ad)} = f_{Ca} \overline{N}_{Ca} \qquad [20]$$

$$a^2_K = \gamma^4_{KCl}\, m^2_K \quad \text{and} \quad a_{Ca} = \gamma^3_{CaCl_2}\, m_{Ca} \qquad [21]$$

where $f$ is the activity coefficient of the cation on the exchanger phase, $\overline{N}$ is the mole fraction of the cation on the exchanger phase, $\gamma$ is the mean activity coefficient of the ions in the solution phase, and m is the molality of the cation in the solution phase. Equation [19] therefore assumes the form:

$$\frac{f^2_K \overline{N}^2_K\, \gamma^3_{CaCl_2}\, m_{Ca}}{f_{Ca}\, \overline{N}_{Ca}\, \gamma^4_{KCl}\, m^2_K} = K_{eq} \qquad [22]$$

or

$$K_{eq} = \frac{f^2_K}{f_{Ca}}\, k_v \qquad [23]$$

where $k_v$ is the Vanselow selectivity coefficient.

Applying the Gibbs-Duhem equation to this binary system and combining with equation [22], one may obtain the following using a mole fraction of unity as the standard state for adsorbed ions (Argersinger *et al.*, 1950):

$$\ln K_{eq} = \int_0^1 \ln k_v\, d\overline{X}_K \qquad [24]$$

where $\overline{X}_K$ is the equivalent fraction of K on the colloid. Integrals are then evaluated graphically by plotting in ln $k_v$ against $\overline{X}_K$. The area under the curve that equals ln $K_{eq}$ can be obtained using the trapezoidal rule.

The standard free energy, $\Delta G°$; enthalpy, $\Delta H°$; and entropy, $\Delta S°$, changes for K exchange are then calculated from equations [25] to [27]:

$$\Delta G° = -RT \ln K_{eq} \qquad [25]$$

$$\ln\left(\frac{K_{eq_2}}{K_{eq_1}}\right) = \frac{-\Delta H°}{R}\left(\frac{1}{T_2} - \frac{1}{T_1}\right) \qquad [26]$$

$$\Delta S° = (\Delta H° - \Delta G°)/T \qquad [27]$$

2. Kinetics Approach

The general derivation of a kinetics model for cation exchange is based on the rate of the reaction in equation [18]. One assumes at the outset that this rate can be expressed as (Denbigh, 1981, Sposito, 1985) as:

$$dE = R_f - R_b \qquad [28]$$

where $dE$ is the extent of the reaction parameter, $R_f$ is the rate of the forward and $R_b$ is the rate of the backward reaction in equation [18]; i.e., the same reaction describes both the forward and the backward processes in the K–Ca reaction.

The individual rates $R_f$ and $R_b$ depend on temperature, pressure, and the concentrations of the species in equation [18]. At equilibrium, the left side of equation [28] will vanish and $R_{fe}/R_{be}$, where $e$ denotes the equilibrium condition, will be some function of temperature ($T$), pressure ($P$), and the equilibrium composition of the exchanger and the aqueous solution phase. Because the activity coefficients of the species in equation [18] have the same kind of functional dependence, it may be inferred that $R_{fe}/R_{be}$ is a function of $T$, $P$, and the species activities (Denbigh, 1981). However, this relationship applies also to the quotient of the right and left sides of equation [22] and, therefore, it may be concluded that, for example,

$$R_{fe}/R_{be} = F\,[(a_{\mathrm{K}(ad)})_e^2\,(a_{\mathrm{Ca}})_e/(a_{\mathrm{Ca}(ad)})_e\,(a_{\mathrm{K}})_e^2 K_{\mathrm{eq}}] \qquad [29]$$

in respect to equation [19].

Equation [29] expresses a general relation between reaction rate and thermodynamics, predicted on equation [28] and the uniqueness of equation [18] as experimental facts of kinetics. If it is assumed further that $F(x) = x$ on the right side of equation [29], then $R_{fe}/R_{be}$ will equal the quotient of the right and left sides of Eq. [22]:

$$R_{fe}/R_{be} = (a_{\mathrm{K}(ad)})_e^2\,(a_{\mathrm{Ca}})_e\,/\,(a_{\mathrm{Ca}(ad)})_e\,(a_{\mathrm{K}})_e^2\,K_{\mathrm{eq}} \qquad [30]$$

Equation [30] is expected whenever $R_f$ and $R_b$ depend on powers of the concentrations of the reactants and products in equation [18] and the power exponents are the stoichiometric coefficients of the four species involved (Denbigh, 1981). Even with this simplification of equation [2], it is necessary to assume that the mechanism of the cation exchange reaction at equilibrium does not change when $R_f \neq R_b$ in order that finite rate data can be used in the application of equation [30].

For the kinetic phase of the study by Ogwada and Sparks (1986a), the following exchange reaction was investigated:

$$\mathrm{Ca}_{1/2}\,\mathrm{X}\,(s) + \mathrm{K}^+\,(aq) = \mathrm{KX}\,(s) + \tfrac{1}{2}\,\mathrm{Ca}^{2+}\,(aq) \qquad [31]$$

One may theorize that the applicability of equation [28] to equation [31] depends on the assumption that the rate is not controlled by diffusion processes. Helfferich (1965, 1983) has noted that if the rate-controlling process is diffusion and not the reaction in equation [31], then no information about $K_{\mathrm{eq}}$ can be derived from an analysis of kinetics data. This hypothesis was tested in the study of Ogwada and Sparks (1986a). Assuming the influence of diffusion is eliminated or practically so, the rates $R_f$ and $R_b$ can be modeled for K–Ca exchange using the equations:

$$R_f = k_a M_s q_{\mathrm{Ca}}\, m_{\mathrm{K}}\ ;\ R_b = k_d M_s q_{\mathrm{K}} m_{\mathrm{Ca}}^{1/2} \qquad [32]$$

where $k_a$ and $k_d$ are adsorption and desorption rate coefficients, respectively, $M_s$ is the mass of the exchanger, $q_i$ ($i$ = Ca or K) is the number of moles of charge of metal $i$ adsorbed by unit mass of the exchanger, and $m_i$ ($i$ = Ca or K) is a molality. The rate of formation of KX (s) now can be expressed by the equation:

$$\frac{dE}{dt} = M_s \frac{dq_{\mathrm{K}}}{dt} = M_s\,[k_a q_{\mathrm{Ca}} m_K - k_d q_{\mathrm{K}} m_{\mathrm{Ca}}^{1/2}] \qquad [33]$$

If the rate of adsorption of K is measured by miscible displacement (Sparks and Jardine, 1981) with $m_{\mathrm{K}}$ maintained constant and $m_{\mathrm{Ca}} = 0$, then equation [33] reduces to equation [34]:

$$\frac{dq_{\mathrm{K}}}{dt} = k'_a m_{\mathrm{K}}(q_T - q_{\mathrm{K}}) \qquad [34]$$

where $q_T = q_{\mathrm{K}} + q_{\mathrm{Ca}}$ is the cation-exchange capacity and $k'_a$ is now an apparent adsorption rate coefficient (Sparks *et al.*, 1980c). Equation [34] is a first-order rate equation with respect to $K^+$ (aq) and KX (s). If the rate of desorption is measured with $m_{\mathrm{Ca}}$ held constant and $m_{\mathrm{K}} = 0$ as in the experiment of Ogwada and Sparks (1986a), equation [33] becomes:

$$\frac{dq_{\mathrm{K}}}{dt} = -k'_d m_{\mathrm{Ca}}^{1/2} q_{\mathrm{K}} \qquad [35]$$

where $k'_d$ is an apparent desorption rate coefficient (Sparks *et al.*, 1980c) that is first-order with respect to KX (s) but of fractional order in respect to $Ca^{2+}$ (aq). Equations [34] and [35] lead to measurement of the coefficients $k'_a$ and $k'_d$ by standard kinetic analysis (Sparks and Jardine, 1981; Sparks, 1985a,b; Sparks and Huang, 1985).

Once $K_{\mathrm{eq}}$ is determined through a kinetic approach, e.g., $k_a/k_d = K_{\mathrm{eq}}$, $\Delta G°$ can be obtained using equation [25].

Using the Arrhenius and van't Hoff equations (Denbigh, 1981), energies of activation for K adsorption and for K desorption ($E_{aa}$ and $E_{ad}$, respectively) can be determined as shown below (Sparks, 1985a,b):

$$d \ln k_a/dT = E_{aa}/RT^2 \qquad [36]$$

and for the desorption kinetic reaction,

$$d \ln k_d/dT = E_{ad}/RT^2 \qquad [37]$$

Substituting,

$$d \ln k_a/dT - d \ln k_d/dT = d \ln K_{\mathrm{eq}}/dt \qquad [38]$$

and from the van't Hoff equation, the enthalpy for K exchange ($\Delta H°$) can be determined:

$$d \ln K_{eq}/dT = \Delta H°/RT^2 \quad [39]$$

or

$$E_{aa} - E_{ad} = \Delta H° \quad [40]$$

The $\Delta S°$ can then be determined using equation [27].

### B. Experimental Results

Ogwada and Sparks (1986a) compared thermodynamic parameters for K–Ca exchange using the equilibrium approach proposed by Argersinger *et al.* (1950) and the kinetic approach outlined earlier on two soils (Table 15). The kinetic approach for determining thermodynamic parameters was investigated using three techniques: a miscible displacement method first proposed by Sparks *et al.* (1980c), a batch, and a vigorously mixed batch. The authors used these three techniques to determine how diffusion affected the comparison between thermodynamic parameters calculated using a kinetic and an equilibrium method.

Ogwada and Sparks (1986a) found that thermodynamic parameters calculated using the equilibrium and kinetic approaches give the same inferences of ion behavior for the two soils studied as indicated by negative $\Delta G°$ and $\Delta H°$ values for both approaches (Table 15). However, except for the vigorously mixed batch technique, the magnitude of the thermodynamic parameters for the two approaches compared poorly (Table 15). The $\Delta G°$ values for both soils at all three temperatures calculated using the vigorously mixed batch technique compared well to those calculated using the equilibrium approach. However, with both the batch and the miscible displacement methods, the comparison was poor. The $\Delta H°$ values using the vigorously mixed kinetic technique and the equilibrium approach compared well, particularly in the Downer soil. The discrepancies are not surprising, because Laudelout *et al.* (1968) even noticed differences in the magnitude of $\Delta H°$ measured by an equilibrium approach and measured directly by calorimetry.

The large differences in the magnitude of the thermodynamic parameters using the equilibrium and the miscible displacement and batch approaches may result from assumptions inherent in both theories. However, the more probable explanation for the discrepancies is the pronounced diffusion that exists particularly in the miscible displacement and batch kinetic methods (Table 15).

The magnitude of the energies of activation for both adsorption and desorption ($E_{aa}$ and $E_{ad}$, respectively) were quite low (Table 15), especially when the miscible displacement and batch techniques were used. These low values suggested that pronounced diffusion-controlled exchange was occurring (Boyd *et al.*, 1947; Mortland and Ellis, 1959; Sparks, 1985a,b).

**Table 15.** Comparison of equilibrium and kinetic approaches for determining thermodynamics of potassium exchange in soils[a]

| | Equilibrium approach | | | |
|---|---|---|---|---|
| Temperature °K | $K_{eq}$ | $\Delta G°$ ($kJ\ mol^{-1}$) | $\Delta H°$ ($kJ\ mol^{-1}$) | $\Delta S°$ ($J\ mol^{-1}\ K^{-1}$) |
| Chester loam | | | | |
| 283 | 9.59 | −5.32 | | −38.72 |
| 298 | 6.77 | −4.74 | | −38.69 |
| 308 | 4.24 | −3.70 | | −40.83 |
| | | | −16.28 | |
| Downer sandy loam | | | | |
| 283 | 9.81 | −5.37 | | −50.72 |
| 298 | 6.43 | −4.61 | | −50.72 |
| 308 | 5.27 | −4.26 | | −50.21 |
| | | | −19.73 | |

| | Kinetic approach | | | | | |
|---|---|---|---|---|---|---|
| | Batch method | | | | | |
| Temperature °K | $K_{eq}$ ($kJ\ mol^{-1}$) | $E_{aa}$ ($kJ\ mol^{-1}$) | $E_{ad}$ ($kJ\ mol^{-1}$) | $\Delta G°$ ($kJ\ mol^{-1}$) | $\Delta H°$ ($kJ\ mol^{-1}$) | $\Delta S°$ ($J\ mol^{-1}\ K^{-1}$) |
| Chester loam | | | | | | |
| 283 | 2.93 | | | −2.53 | | −16.80 |
| 298 | 2.62 | | | −2.39 | | −16.52 |
| 308 | 2.27 | | | −2.10 | | −16.83 |
| | | 10.69 | 18.41 | | −7.72 | |
| Downer sandy loam | | | | | | |
| 283 | 2.49 | | | −2.15 | | −36.73 |
| 298 | 2.76 | | | −2.52 | | −33.64 |
| 308 | 1.49 | | | −1.00 | | −37.45 |
| | | 14.85 | 30.29 | | −15.44 | |

For example, Mortland and Ellis (1959) found that the $E_{ad}$ for K release from vermiculite, was 14.85 kJ mol$^{-1}$ and they concluded that film diffusion was the rate-limiting step. However, Ogwada and Sparks (1986a) found that when a vigorously mixed batch technique was used, the $E_{aa}$ and $E_{ad}$ values were increased, suggesting the effect of diffusion was being diminished significantly (Table 15).

Helfferich (1965, 1983) noted that if the rate-controlling process were diffusion and not the reaction in equation [31], then no information about equilibrium could be derived from kinetic analyses. When diffusion is sig-

**Table 15.** (*Continued*)

| Temperature °K | $K_{eq}$ (kJ mol$^{-1}$) | $E_{aa}$ (kJ mol$^{-1}$) | $E_{ad}$ (kJ mol$^{-1}$) | ΔG° (kJ mol$^{-1}$) | ΔH° (kJ mol$^{-1}$) | ΔS° (J mol$^{-1}$ K$^{-1}$) |
|---|---|---|---|---|---|---|
| | Kinetic approach | | | | | |
| | Miscible displacement method | | | | | |
| Chester loam | | | | | | |
| 283 | 1.55 | | | −1.03 | | −9.14 |
| 298 | 1.38 | | | −0.80 | | −9.58 |
| 308 | 1.33 | | | −0.74 | | −9.44 |
| | | 7.42 | 11.87 | | −4.45 | |
| Downer sandy loam | | | | | | |
| 283 | 2.16 | | | −1.81 | | −12.09 |
| 298 | 1.97 | | | −1.68 | | −12.22 |
| 308 | 1.82 | | | −1.53 | | −12.15 |
| | | 7.42 | 12.47 | | −5.05 | |

| Temperature °K | $K_{eq}$ (kJ mol$^{-1}$) | $E_{aa}$ (kJ mol$^{-1}$) | $E_{ad}$ (kJ mol$^{-1}$) | ΔG° (kJ mol$^{-1}$) | ΔH° (kJ mol$^{-1}$) | ΔS° (J mol$^{-1}$ K$^{-1}$) |
|---|---|---|---|---|---|---|
| | Kinetic approach | | | | | |
| | Vigorously mixed batch method | | | | | |
| Chester loam | | | | | | |
| 283 | 5.45 | | | −3.99 | | −16.69 |
| 298 | 4.66 | | | −3.82 | | −15.96 |
| 308 | 4.01 | | | −3.57 | | −16.24 |
| | | 32.96 | 42.10 | | −9.14 | |
| Downer sandy loam | | | | | | |
| 283 | 11.55 | | | −5.76 | | −41.81 |
| 298 | 7.53 | | | −5.00 | | −42.24 |
| 308 | 6.19 | | | −4.67 | | −42.00 |
| | | 22.39 | 40.92 | | −18.53 | |

[a]From Ogwada and Sparks (1986b).

nificantly reduced, as with the vigorously mixed batch technique, it appears from the data of Ogwada and Sparks (1986a) that a kinetic approach can be used successfully to gather thermodynamic information about a soil system (Table 15). Under the extreme conditions of rapid mass transfer as found with the vigorously mixed batch system, the influence of diffusion is significantly reduced and the rate coefficients obtained approximate that of the reaction step.

The use of a kinetic approach to determine thermodynamic parameters would have many advantages over a traditional equilibrium approach. The most significant advantage would be the ability not only to gain thermodynamic parameters for ion exchange, but also to gather information about rate processes and mechanisms. This type of information is integral to modeling the movement of K in soils.

## XI. Summary and Future Research Needs

Kinetics of ionic reactions in soils and clay minerals are both an exciting and a complicated area of research. This is certainly true for the dynamics of soil potassium. The dynamic reactions that exist between the four forms of potassium are controlled by a number of factors. Perhaps the most important is the type and quantity of organic and inorganic colloids present in the soil. Potassium dynamics can be interpreted using a number of equations and methodologies that are discussed in this contribution.

Although we have learned a great deal about potassium dynamics in soils, it is fair to state that many unanswered questions remain. The role that the sand and silt fractions play in supplying K to plants should be studied further. Additionally, the rate-limiting steps involved in potassium dynamics are not well understood. Finally, if we are to predict fully the fate of applied K in soils, which will assist in increasing crop yields and uptake efficiency, we must derive predictive kinetic models.

### Acknowledgments

The author thanks the Potash and Phosphate Institute for their partial financial support of this review. Gratitude is also expressed to Ms. Muriel Ryder and Mr. G.J. Hendricks for their assistance in revising and typing the manuscript.

### References

Adams, F. 1974. Soil solution. In: E.W. Carson (ed.) *The Plant Root and Its Environment*. University Press of Virginia, Charlottesville, VA, pp. 441–448.

Adams, F., and Z. Rawajfih. 1977. Basaluminite and alunite: A possible cause of sulfate retention by acid soils. *Soil Sci. Soc. Am. J.* 41:686–692.

Aharoni, C., and M. Ungarish. 1976. Kinetics of activated chemisorption. Part I. The non-Elovichian part of the isotherm. *J. Chem. Soc., Faraday Trans.* 72:400–408.

Ahrens, L.H. 1965. *Distribution of Elements in Our Planet*. McGraw-Hill Book Co., New York.

Alexiades, C.A., and M.L. Jackson. 1966. Quantitative clay mineralogical analysis of soils and sediments. *Clays Clay Miner.* 14:35–52.

Alexiades, C.A., N.A. Polyzopoulos, N.A. Koroxenides, and G.S. Axaris. 1973. High trioctahedral vermiculite content in the sand, silt, and clay fractions of a Gray Brown Podzolic soil in Greece. *Soil Sci.* 116:363–375.

Ames, L.L. 1960. The cation sieve properties of clinoptilite. *Am. Mineral.* 45:689–700.

Ardakani, M.S., and A.D. McLaren. 1977. Absence of local equilibrium during ammonium transport in a soil column. *Soil Sci. Soc. Am. J.* 41:877–879.

Argersinger, W.J., Jr., A.W. Davidson, and O.D. Bonner. 1950. Thermodynamics and ion exchange phenomena. *Trans. Kansas Acad. Sci.* 53:404–410.

Arnold, P.W. 1958. Potassium uptake by cation-exchange resins from soils and minerals. *Nature* (London) 182:1594–1595.

Atkinson, R.J., F.J. Hingston, A.M. Posner, and J.P. Quirk. 1970. Elovich equation for the kinetics of isotope exchange reactions at solid-liquid interfaces. *Nature* (London) 226:148–149.

Atschuler, Z.S., E.J. Dwornik, and H. Kramer. 1963. Transformation of montmorillonite to kaolinite during weathering. *Science* (Washington, DC) 141:148–152.

Ayres, A.S., M. Takahashi, and Y. Kanehiro. 1947. Conversion of nonexchangeable potassium to exchangeable forms in a Hawaii soil. *Soil Sci. Soc. Am. Proc.* 11:175–181.

Babcock, K.L. 1963. Theory of the chemical properties of soil colloidal systems at equilibrium. *Hilgardia* 34:417–542.

Bailey, S.W. 1980. Structures of layer silicates. In: G.W. Brindley and G. Brown (eds.) *Crystal Structure of Clay Minerals and Their x-Ray Identification.* Monogr. No. 5, Mineralogical Society, London.

Barber, R.G. 1979. Potassium fixation in some Kenyan soils. *J. Soil Sci.* 30:785–792.

Barber, R.G., and B.C. Mathews. 1962. Release of nonexchangeable soil potassium by resin equilibration and its significance for crop growth. *Can. J. Soil Sci.* 42:266–272.

Barnhisel, R.I. 1977. Chlorites and hydroxy-interlayered vermiculite and smectite. In: J.B. Dixon and S.B. Weed (eds.) *Minerals in soil environments.* Soil Science Society of America, Madison, WI, pp. 331–356.

Barnhisel, R.I., and P.M. Bertsch. 1982. Aluminum. In: A.L. Page (ed.) *Methods of Soil Analysis,* Part 2. *Chemical and Microbiological Properties.* American Society of Agronomy and Soil Science Society of America, Madison, WI, pp. 275–300.

Barrer, R.M., J.A. Davis, and L.V.C. Rees. 1969. Thermodynamics and thermochemistry of cation exchange in chabazite. *J. Inorg. Nucl. Chem.* 31:219–232.

Barrer, R.M., and B. Munday. 1971. Cation exchange reactions of a sedimentary phillipsite. *J. Chem. Soc. Am.* 24:2904–2909.

Barrow, N.J. 1983. A discussion of the methods for measuring the rate of reaction between soil and phosphate. *Fertil. Res.* 40:51.

Barrow, N.J., and T.C. Shaw. 1979. Effects of solution: soil ratio and vigour of shaking on the rate of phosphate adsorption by soil. *Soil Sci.* 119:167–177.

Barth, T.W.F. 1969. *Feldspars.* Wiley-Interscience, New York.

Bassett, W.A. 1959. The origin of the vermiculite deposit at Libby, Montana. *Am. Mineral.* 44:282–299.

Bassett, W.A. 1960. Rate of hydroxyl orientation in mica alteration. *Geol. Soc. Am. Bull.* 71:449–456.

Berkheiser, V.E., J.J. Street, P.S.C. Rao, and T.L. Yuan. 1980. Partitioning of inorganic orthophosphate in soil-water systems. *CRC Crit. Rev. Environ. Contr.* 10:179–224.

Berner, R.A. 1983. Kinetics of weathering and diagenesis. In: A.C. Lasaga and R.J. Kirkpatrick (eds.) *Kinetics of Geochemical Processes*. Vol. 8. *Reviews in Mineralogy*. Mineralogical Society of America, Washington, DC, pp. 111–134.

Berner, R.A., and G.R. Holdren, Jr. 1979. Mechanism of feldspar weathering II. Observations of feldspars from soils. *Geochim. Cosmochim. Acta* 43:1173–1186.

Bertsch, P.M., and G.W. Thomas. 1985. Potassium status of temperate region soils. In: R.E. Munson (ed.) *Potassium in Agriculture*. American Society of Agronomy, Crop Science Society of America, and Soil Science Society of America, Madison, WI, pp. 131–162.

Bolt, G.H., M.E. Sumner, and A. Kamphorst. 1963. A study between three categories of potassium in an illitic soil. *Soil Sci. Soc. Am. Proc.* 27:294–299.

Borchardt, G.A. 1977. Montmorillonite and other smectite minerals. In: J.B. Dixon and S.B. Weed (eds.) Minerals in soil environments. Soil Science Society of America, Madison, WI, pp. 293–330.

Boyd, G.E., A.W. Adamson, and L.S. Myers, Jr. 1947. The exchange adsorption of ions from aqueous solutions by organic zeolites: II. Kinetics. *J. Am. Chem. Soc.* 69:2836–2848.

Boyle, J.R., G.K. Voight, and B.L. Sawhney. 1967. Biotite flakes: Alteration by chemical and biological treatment. *Science* (Washington, DC) 155:193–195.

Burns, A.F., and S.A. Barber. 1961. Effect of temperature and moisture on exchangeable potassium. *Soil Sci. Soc. Am. Proc.* 25:349–352.

Busenberg, E., and C.V. Clemency. 1976. The dissolution kinetics of feldspars at 25°C and 1 atm $CO_2$ partial pressure. *Geochim. Cosmochim. Acta* 40:41–50.

Carski, T.H., and D.L. Sparks. 1985. A modified miscible displacement technique for investigating adsorption-desorption kinetics in soils. *Soil Sci. Soc. Am. J.* 49:1114–1116.

Chien, S.H., and W.R. Clayton. 1980. Application of Elovich equation to the kinetics of phosphate release and sorption in soils. *Soil Sci. Soc. Am. J.* 44:265–268.

Chien, S.H., W.R. Clayton, and G.H. McClellan. 1980. Kinetics of dissolution of phosphate rocks in soils. *Soil Sci. Soc. Am. J.* 44:260–264.

Chute, J.H., and J.P. Quirk. 1967. Diffusion of potassium from mica-like materials. *Nature* (London) 213:1156–1157.

Coffman, C.B., and D.S. Fanning. 1974. "Vermiculite" determination on whole soils by cation exchange capacity methods. *Clays Clay Miner.* 22:271–283.

Conyers, E.S., and E.O. McLean. 1969. Plant uptake and chemical extractions for potassium release characteristics of soils. *Soil Sci. Soc. Am. Proc.* 33:226–230.

Cooke, G.W. 1979. Some priorities for British soil science. *J. Soil Sci.* 30:187–213.

Cooke, G.W., and P.A. Gething. 1978. Changing concepts on the use of potash. *Congr. Int. Potash Inst.* 11:361–405.

Denbigh, K. 1981. *The Principles of Chemical Equilibrium*. Cambridge University Press, Cambridge.

Dibble, W.E., Jr., and W.A. Tiller. 1981. Non-equilibrium water/rock interactions. I. Model for interface-controlled reactions. *Geochim. Cosmochim. Acta* 45:79–92.

Dixon, J.B., and S.B. Weed. 1977. *Minerals in Soil Environments*. Soil Science Society of America, Madison, WI.

Doll, A.C., and R.E. Lucas. 1973. Testing soils for potassium, calcium, and magnesium. In: L.M. Walsh and J.D. Beaton (eds.) *Soil Testing and Plant Analysis*. Soil Science Society of America, Madison, WI, pp. 133–151.

Douglas, L.A. 1977. Vermiculites. In: J.B. Dixon and S.B. Weed (eds.) *Minerals in Soil Environments*. Soil Science Society of America, Madison, WI, pp. 259–292.

Evans, C.M. 1983. Kinetics and mechanism of boron adsorption and desorption in Atlantic Coastal Plain soils. M.S. Thesis, University of Delaware, Newark, DE.

Eyring, H., S.H. Lin, and S.M. Lin. 1980. *Basic Chemical Kinetics*. John Wiley, New York.

Fanning, D.S., and V.Z. Keramidas. 1977. Micas. In: J.B. Dixon and S.B. Weed (eds.) *Minerals in Soil Environments*. Soil Science Society of America, Madison, WI, pp. 195–258.

Feigenbaum, S., and R. Levy. 1977. Potassium release in some saline soils of Israel. *Geoderma* 19:159–169.

Feigenbaum S., and I. Shainberg, 1975. Dissolution of illite—a possible mechanism of potassium release. *Soil Sci. Soc. Am. J.* 39:985–990.

Feigenbaum, S., R. Edelstein, and I. Shainberg. 1981. Release rate of potassium and structural cations from micas to ion exchangers in dilute solutions. *Soil Sci. Soc. Am. J.* 45:501–506.

Frost, A.A., and R.G. Pearson. 1961. *Kinetics and Mechanism*. John Wiley and Sons, New York.

Garman, W.L. 1957. Potassium release characteristics of several soils from Ohio and New York. *Soil Sci. Soc. Am. Proc.* 21:52–58.

Gast, R.G. 1977. Surface and colloid chemistry. In: J.B. Dixon and S.B. Weed (eds.) Minerals in soil environments. *Soil Science Society of America,* Madison, WI, pp. 27–73.

Gedroiz, K.K. 1914. Colloidal chemistry as related to soil science: II. Rapidity of reaction exchange in the soil, colloidal condition of the soil saturated with various bases and the indicator method of determining the colloidal content of the soil. *Zh. Opytn. Agron.* 15:181–208.

Gilligan, G.M. 1936. The effect of fertilizers and cropping upon the nature and amount of electrodialyzable bases in the soil with particular reference to potassium. Bull. 200 Delaware Agricultural Experiment Station, Newark, Delaware.

Glasstone, S., K.J. Laidler, and H. Eyring. 1941. *The Theory of Rate Processes*. McGraw-Hill, New York.

Goulding, K.W.T. 1980. The thermodynamics of ion exchange adsorption in soils and soil clay minerals. Ph.D. Diss. University of London, London.

Goulding, K.W.T. 1981. Potassium retention and release in Rothamsted and Saxmundhum soils. *J. Sci. Food Agric.* 32:617–670.

Goulding, K.W.T. 1983. Thermodynamics and potassium exchange in soils and clay minerals. *Adv. Agron.* 36:215–283.

Goulding, K.W.T., and O. Talibudeen. 1979. Potassium reserves in a sandy clay soil from the Saxmundhum experiment: Kinetics and equilibrium thermodynamics. *J. Soil Sci.* 30:291–302.

Grimme, H. 1979. Factors controlling potassium availability to plants. In: A. Saurat and M.M. El-Fouly (eds.) *Role of K in crop production.* Proc. Int. Potassium Workshop, Cairo, Egypt, pp. 33–47.

Grimme, H. 1980. *The evaluation of Soil K Status by means of Soil Testing.* Potassium Workshop, Ibadan, Nigeria, pp. 23–40.

Haagsma, T. 1961. The release of nonexchangeable soil potassium to cation exchange resins as influenced by temperature and moisture. M.S. Thesis, University of Toronto, Toronto, Canada.

Haagsma, T., and M.H. Miller. 1963. The release of nonexchangeable soil potassium to cation exchange resins as influenced by temperature, moisture and exchanging ion. *Soil Sci. Soc. Am. Proc.* 27:153–156.

Harter, R.D., and R.G. Lehmann. 1983. Use of kinetics for the study of exchange reactions in soils. *Soil Sci. Soc. Am. J.* 47:666–669.

Havlin, J.L., and D.G. Westfall. 1985. Potassium release kinetics and plant response in calcareous soils. *Soil Sci. Soc. Am. J.* 49:366–370.

Havlin, J.L., D.G. Westfall, and S.R. Olsen. 1985. Mathematical models for potassium release kinetics in calcareous soils. *Soil Sci. Soc. Am. J.* 49:371–376.

Heier, K.S. 1962. Trace elements in feldspars, a review. *Nor. Geol. Tidski.* 42 (Feldspar Vol.): 415–455.

Helfferich, F. 1962. *Ion Exchange.* McGraw-Hill Book Co., New York.

Helfferich, F. 1965. Ion exchange kinetics. V. Ion exchange accompanied by reactions. *J. Phys. Chem.* 69:178.

Helfferich, F. 1983. Ion exchange kinetics-Evolution of a theory. In: L. Liberti and F. Helfferich (eds.) *Mass Transfer and Kinetics of Ion Exchange.* Martinus Nijhoff, The Hague, pp. 157–179.

Helgeson, H.C. 1971. Kinetics of mass transfer among silicates and aqueous solutions. *Geochim. Cosmochim. Acta* 35:421–469.

Hendricks, G.J., D.L. Sparks, and D.R. Parker. 1985. Effect of timing of potassium applications on leaching and corn yields in Delaware soils in an intensive management system. *Agron. Abstr.*, American Society of Agronomy, Madison, WI, p. 25.

Hissink, D.J. 1924. Base exchange in soils. *Trans. Faraday Soc.* 20:551–566.

Holdren, G.R., Jr., and R.A. Berner. 1979. Mechanism of feldspar weathering. I. Experimental studies. *Geochim. Cosmochim. Acta* 43:1161–1171.

Holmes, A. 1965. *Principles of Physical Geology.* Ronald Press, New York.

Hossner, L.R. 1966. Release of magnesium by leaching from vermiculite, mica and prochlorite. Ph.D. Thesis. Michigan State University, Univ. Microfilms. Ann Arbor, Mich. (*Diss. Abstr.* 26:4140).

Huang, P.M. 1979. Soil potassium reserves in relation to crop response to potash. Proceedings of the Workshops, Potash and Phosphate Institute of Canada, Saskatoon, Saskatchewan, Canada.

Huang, P.M. 1980. Adsorption processes in soils. In: O. Hutzinger (ed.) *The Handbook of Environmental Chemistry,* Vol. 2, Part A. *Reactions and Processes.* Springer-Verlag, New York, pp. 47–59.

Huang, P.M., and S.Y. Lee. 1969. Effect of drainage on weathering transformations of mineral colloids of some Canadian Prairie soils. *Proc. Int. Clay Conf. Tokyo* 1:541–551.

Huang, P.M., L.S. Crossan, and D.A. Rennie. 1968. Chemical dynamics of K release from potassium minerals common in soils. *Trans. 9th Int. Congr. Soil Sci.* 2:705–612.

Huff, W.D. 1972. Morphological effects as a result of potassium depletion. *Clays Clay Miner.* 20:295–301.

Iskandar, R.S. 1981. *Modeling Wastewater Renovation.* John Wiley and Sons, New York.

Jackson, M.L. 1959. Frequency distribution of clay minerals in major great soil groups as related to the factors of soil formation. *Clays Clay Miner.* 6:133–143.

Jackson, M.L. 1964. Chemical composition of soils. In: F.E. Bear (ed.) *Chemistry of the Soil.* Van Nostrand Reinhold, New York, pp. 71–141.

Jackson, M.L., and G.D. Sherman. 1953. Chemical weathering of minerals in soils. *Adv. Agron.* 5:219–318.

Jackson, M.L., Y. Hseung, R.B. Corey, E.J. Evans, and R.C. Vanden Heuvel. 1952. Weathering of clay-size minerals in soils and sediments: II. Chemical weathering of layer silicates. *Soil Sci. Soc. Am. Proc.* 16:3–6.

Jardine, P.M., and D.L. Sparks. 1984a. Potassium-calcium exchange in a multireactive soil system: I. Kinetics. *Soil Sci. Soc. Am. J.* 47:39–45.

Jardine, P.M., and D.L. Sparks. 1984b. Potassium-calcium exchange in a multireactive soil system: II. Thermodynamics. *Soil Sci. Soc. Am. J.* 47:45–50.

Jensen, H.E. 1973. Potassium-calcium exchange equilibria on a montmorillonite and a kaolinite clay: I. A test on the Argersinger thermodynamic approach. *Agrochimica* 17:181–190.

Jensen, J.R. 1984. Potassium dynamics in soil. *Soil Sci.* 138:285–293.

Kato, Y. 1965. Mineralogical study of weathering products of granodiosite at Shinshiro City: IV. Mineralogical compositions of silt and clay fractions. *Soil Sci. Plant Nutrit.* (Tokyo) 11:16–27.

Keay, J., and A. Wild. 1961. The kinetics of cation exchange in vermiculite. *Soil Sci.* 92:49–54.

Keller, W.D. 1955. *The Principles of Chemical Weathering.* Lucas Brothers, Columbia, Missouri.

Kelley, W.P. 1948. *Cation Exchange in Soils.* American Chemical Society Monogr. 109. Van Nostrand Reinhold, New York.

Kittrick, J.A. 1966. Forces involved in ion fixation by vermiculite. *Soil Sci. Soc. Am. Proc.* 30:801–803.

Kittrick, J.A. 1973. Mica-derived vermiculites as unstable intermediates. *Clays Clay Miner.* 21:479–488.

Komarneni, S. 1978. Cesium sorption and desorption behavior of kaolinites. *Soil Sci. Soc. Am. J.* 42:531–533.

Kottgen, P., and R. Diehl. 1929. Ueber die anwendung der dialyse und electroultrafiltration zur bestimmung des nahrstaffbedurfnisses des bodens. *Z. Pflanzenernahr, Dung-u. Bodenk.* 12:65–105.

Kozak, L.M., and P.M. Huang. 1971. Adsorption of hydroxy-Al by certain phyllosilicates and its relation to K/Ca cation exchange selectivity. *Clays Clay Miner.* 19:95–102.

Kyle, J.H., A.M. Posner, and J.P. Quirk. 1975. Kinetics of isotopic exchange of phosphate adsorbed on gibbsite. *J. Soil Sci.* 26:32–42.

Laidler, K.J. 1965. *Chemical Kinetics.* McGraw-Hill Book Co., New York.

Laudelout, H., R. van Bladel, G.H. Bolt, and A.L. Page. 1968. Thermodynamics of heterovalent cation exchange reactions in a montmorillonite clay. *Trans. Faraday Soc.* 64:1477–1488.

Lee, S.Y., M.L. Jackson, and J.L. Brown. 1975a. Micaceous occlusions in kaolinite observed by ultramicrotomy and high resolution electron microscopy. *Clays Clay Miner.* 23:125–129.

Lee, S.Y., M.L. Jackson, and J.L. Brown. 1975b. Micaceous vermiculite, glauconite, and mixed-layered kaolinite-montmorillonite examination by ultramicrotomy and high resolution electron microscopy. *Soil Sci. Soc. Am. Proc.* 39:793–800.

Low, M.J.D. 1960. Kinetics of chemisorption of gases on solids. *Chem. Rev.* 60:265–312.

MacLean, A.J. 1961. Potassium-supplying power of some Canadian soils. *Can. J. Soil Sci.* 41:196–206.

Malavolta, E. 1985. Potassium status of tropical and subtropical region soils. In: R.E. Munson (ed.) *Potassium in Agriculture*. American Society of Agronomy, Crop Science Society of America, and Soil Science Society of America, Madison, WI, pp. 163–200.

Malcom, R.L., and V.C. Kennedy. 1969. Rate of cation exchange on clay minerals as determined by specific ion electrode techniques. *Soil Sci. Soc. Am. Proc.* 33:245–253.

Malquori, A., G. Ristone, and V. Vidrich. 1975. Biological weathering of potassium silicate: I. Biotite. *Potash Rev.* 3:1–6.

Mansell, R.S., and H.M. Selim. 1981. Mathematical models for predicting reactions and transport of phosphorus applied to soils. In: R.S. Iskandar (ed.) *Modeling Wastewater Renovation*. John Wiley and Sons, New York, pp. 600–646.

Martin, H.W., and D.L. Sparks. 1983. Kinetics of nonexchangeable potassium release from two Coastal Plain soils. *Soil Sci. Soc. Am. J.* 47:883–887.

Martin, H.W., and D.L. Sparks. 1985. On the behavior of nonexchangeable potassium in soils. *Commun. Soil Sci. Plant Anal.* 16:133–162.

Mengel, K. 1985. Dynamics and availability of major nutrients in soils. *Adv. Soil Sci.* 1:65–131.

Milford, M.H., and M.L. Jackson. 1960. Exchangeable potassium as affected by mica specific surface in some soils of north Central United States. *Soil Sci. Soc. Am. Proc.* 30:735–739.

Mirecki, J., M.R. Noll, and D.L. Sparks. 1985. Sand mineralogy of Delaware Coastal Plain soils: Its role in supplying potassium. *Agron. Abstr.* American Society of Agronomy, Madison, WI, p. 232.

Mortland, M.M. 1958. Kinetics of potassium release from biotite. *Soil Sci. Soc. Am. Proc.* 22:503–508.

Mortland, M.M. 1961. The dynamic character of potassium release and fixation. *Soil Sci.* 91:11–13.

Mortland, M.M., and B.G. Ellis. 1959. Release of fixed potassium as a diffusion-controlled process. *Soil Sci. Soc. Am. Proc.* 23:363–364.

Mortland, M.M., K. Lawton, and G. Uehara. 1956. Alteration of biotite to vermiculite by plant growth. *Soil Sci.* 82:477–481.

Murali, V., and L.A.G. Aylmore. 1981. A convective-dispersive-adsorptive flow model for solute transport in soils: I. Model description and some simulations. *Austr. J. Soil Res.* 19:23–29.

Murali, V., and L.A.G. Aylmore. 1983a. Competitive adsorption during solute transport in soils: I. Mathematical models. *Soil Sci.* 135:143–150.

Murali, V., and L.A.G. Aylmore. 1983b. Competitive adsorption during solute transport in soils: 3. A review of experimental evidence of competitive adsorption and an evaluation of simple competition models. *Soil Sci.* 136:279–290.

Naylor, D.V., and R. Overstreet. 1969. Sodium-calcium exchange behavior in organic soils. *Soil Sci. Soc. Am. Proc.* 33:848–851.

Nemeth, K. 1971a. Die charakterisierung des K-haushaltes von boden mittels K-desorption skurven. *Geoderma* 5:99–101.

Nemeth, K. 1971b. Moglichkeiten zur bestimmung massgeblicher faktoren der bodenfruchtbarkeit mittels electro-ultrafiltration (EUF). *Landw. Farsch. Sonderh.* 26:192–198.

Nemeth, K. 1972. The determination of desorption and solubility rates of nutrients in the soil by means of electroultrafiltration (EUF). *Proc. Collog. Int. Potash Inst.* 9:171–180.

Nemeth, K. 1976. The determination of effective and potential availability of nutrients in the soil by electroultrafiltration. *Appl. Sci. Devel.* 8:89–111.

Nemeth, K. 1979. The availability of nutrients in the soil as determined by electroultrafiltration (EUF). *Adv. Agron.* 31:155–188.

Norrish, K. 1967. Clay mineralogy and sources of natural potassium. In: *Potassium in Southeastern Australia*. Australian Institute of Agricultural Sciences, pp. 6–8.

Norrish, K. 1973. Factors in the weathering of mica to vermiculite. In: J.M. Serratosa (ed.) *1972 Proceedings of the 4th International Clay Conference*. pp. 417–432.

Nye, P.H. 1981. Changes of pH across the rhizosphere induced by roots. *Plant Soil* 61:7–26.

Ogwada, R.A., and D.L. Sparks. 1986a. A critical evaluation on the use of kinetics for determining thermodynamics of ion exchange in soils. *Soil Sci. Soc. Am. J.* 50:300–305.

Ogwada, R.A., and D.L. Sparks. 1986b. Kinetics of ion exchange in clay minerals and soil: I. Evaluation of methods. *Soil Sci. Soc. Am. J.* 50:1158–1162.

Onken, A.B., and R.L. Matheson. 1982. Dissolution rate of EDTA-extractable phosphate from soils. *Soil Sci. Soc. Am. J.* 46:276–279.

Peech, M., and R. Bradfield. 1934. The effect of lime and neutral calcium salts upon the solubility of soil potassium. *Amer. Soil Surv. Assoc. Bull.* 15:101–106.

Peterson, D.L., H. Helfferich, and G.C. Blytas. 1965. Sorption and ion exchange in sedimentary zeolites. *J. Phys. Chem. Solids* 26:835–848.

Petrovic, R. 1976. Rate control in feldspar dissolution II. The protective effect of precipitates. *Geochim. Cosmochim. Acta* 4:1509–1522.

Petrovic, R., R.A. Berner, and Goldhaber, M.B. 1976. Rate control in dissolution of alkali feldspars I. Study of residual feldspar grains by x-ray photoelectron spectroscopy. *Geochim. Cosmochim. Acta* 40:537–548.

Pratt, P.F. 1951. Potassium removed from Iowa soils by greenhouse and laboratory procedures. *Soil Sci.* 72:107–117.

Pratt, P.F. 1965. Potassium. In: C.A. Black (ed.) *Methods of Soil Analysis*. Part 2. *Chemical and Microbiological Properties*. American Society of Agronomy, Madison, WI, pp. 1023–1031.

Pratt, P.F., and H.H. Morse. 1954. Potassium release from exchangeable and nonexchangeable forms in Ohio soils. Res. Bull. 747, Ohio Agricultural Experiment Station, Columbus, Ohio.

Quirk, J.P., and J.H. Chute. 1968. Potassium release from mica-like clay minerals. *Trans. 9th Int. Congr. Soil Sci.* 2:671.
Rankama, K., and Th.G. Sahama. 1950. *Geochemistry*. University of Chicago Press, Chicago.
Rao, P.S.C., and J.M. Davidson. 1978. Nonequilibrium conditions for ammonium-adsorption-desorption during flow in soils. *Soil Sci. Soc. Am. J.* 42:668.
Rao, P.S.C., and R.E. Jessup. 1982. Development and verification of simulation models for describing pesticide dynamics in soils. *Ecol. Model.* 16:67–75.
Rasmussen, K. 1972. Potash in feldspars. *Proc. Colloq. Int. Potash Inst.* 9:57–60.
Rausell-Colom, J.A., T.R. Sweetman, L.B. Wells, and K. Norrish. 1965. Studies in the artificial weathering of micas. In: E.G. Hallsworth and D.V. Crawford (eds.) *Experimental Pedology*. Butterworths, London, pp. 40–70.
Reed, M.G., and A.D. Scott. 1962. Kinetics of potassium release from biotite and muscovite in sodium tetraphenylboron solutions. *Soil Sci. Soc. Am. Proc.* 26:437–440.
Reitemeier, R.F. 1951. The chemistry of soil potassium. *Adv. Agron.* 3:113–164.
Rich, C.I. 1958. Muscovite weathering in a soil developed in the Virginia piedmont. *Clays Clay Miner.* 5:203–212.
Rich, C.I. 1960. Aluminum in interlayers of vermiculite. *Soil Sci. Soc. Am. Proc.* 24:26–32.
Rich, C.I. 1964. Effect of cation size and pH on potassium exchange in Nason soil. *Soil Sci.* 98:100–106.
Rich, C.I. 1968. Mineralogy of soil potassium. In: V.J. Kilmer, S.E. Younts, and N.C. Brady, (eds.) *The Role of Potassium in Agriculture*. American Society of Agronomy, Crop Science Society of America, and Soil Science Society of America, Madison, WI, pp. 79–96.
Rich, C.I. 1972. Potassium in minerals. *Proc. Colloq. Int. Potash Inst.* 9:15–31.
Rich, C.I., and W.R. Black. 1964. Potassium exchange as affected by cation size, pH, and mineral structure. *Soil Sci.* 97:384–390.
Rich, C.I., and S.S. Obenshain. 1955. Chemical and clay mineral properties of a red-yellow podzolic soil derived from mica schist. *Soil Sci. Soc. Am. Proc.* 19:334–339.
Ross, G.J. 1971. Relation of potassium exchange and fixation to degree of weathering and organic matter content in micaceous clays of podzol soils. *Clays Clay Miner.* 9:167–174.
Russell, E.W. 1961. *Soil Conditions and Plant Growth*. Longmans, London.
Sadusky, M.C., and D.L. Sparks. 1985. Dynamics of potassium release from the sand fractions of Coastal Plain soils. *Agron. Abstr.* American Society of Agronomy, Madison, WI, p. 151.
Salomon, M., and J.B. Smith. 1957. A comparison of methods for determining extractable soil potassium in fertilizer test plots. *Soil Sci. Soc. Am. Proc.* 21:222.
Sawhney, B.L. 1966. Kinetics of cesium sorption by clay minerals. *Soil Sci. Soc. Am. Proc.* 30:565–569.
Sawhney, B.L., and G.K. Voight. 1969. Chemical and biological weathering in vermiculite from Transvaal. *Soil Sci. Soc. Am. Proc.* 33:625–629.
Schmitz, G.W., and P.F. Pratt. 1953. Exchangeable and nonexchangeable potassium as indexes to yield increases and potassium adsorption by corn in the greenhouse. *Soil Sci.* 76:345–353.

Schofield, K.K., and H.R. Samson. 1954. Flocculation of kaolinite due to the attraction of oppositely charged crystal faces. *Disc. Faraday Soc.* 18:135–145.

Schroeder, D. 1974. Relationships between soil potassium and the potassium nutrition of plants. *Proc. Congr. Int. Potash Inst.* 10:33–63.

Schroeder, D. 1979. Structure and weathering of potassium containing minerals. *Proc. Congr. Int. Potash Inst.* 11:43–63.

Schuffelen, A.C. 1972. The cation exchange system of the soil. *Proc. Colloq. Int. Potash Inst.* 9:75–88.

Scott, A.D. 1968. Effect of particle size on interlayer potassium exchange in micas. *Trans. Int. Congr. Soil Sci.* 2:649–660.

Scott, A.D., and M.G. Reed. 1962. Chemical extraction of potassium from soils and micaceous minerals with solution containing sodium tetraphenylboron: II. Biotite. *Soil Sci. Soc. Am. Proc.* 26:41–45.

Scott, A.D., and S.J. Smith. 1966. Susceptibility of interlayer potassium in micas to exchange with sodium. *Clays Clay Miner. Conf., 14th Proc.* 26:69–81.

Scott, A.D., and S.J. Smith. 1967. Visible changes in macro mica particles that occur with potassium depletion. *Clays Clay Miner.* 15:367–373.

Scott, A.D., A.P. Edwards, and J.M. Bremner. 1960. Removal of fixed ammonium from clay minerals by cation exchange resins. *Nature* (London) 185:792.

Selim, H.M., P. Kanchanasut, R.S. Mansell, L.W. Zelazny, and J.M. Davidson. 1975. Phosphorus and chloride movement in a Spodosol. *Soil Crop Sci. Soc. Fla. Proc.* 34:18–23.

Selim, H.M., R.S. Mansell, and L.W. Zelazny. 1976. Modeling reactions and transport of potassium in soils. *Soil Sci.* 122:77–84.

Sivasubramaniam, S., and O. Talibudeen. 1972. Potassium-aluminum exchange in acid soils: I. Kinetics. *J. Soil Sci.* 23:163–176.

Smith, J.V. 1974. *Feldspar Minerals: Crystal Structure and Physical Properties*, Vol. 1; *Chemical and Textural Properties*, Vol. 2. Springer-Verlag, New York.

Smith, S.J., and A.D. Scott. 1966. Extractable potassium in Grundite illite: I, Method of extraction. *Soil Sci.* 102:115–122.

Soil Survey Staff. 1975. Soil taxonomy: A basic system of soil classification for making and interpreting soil surveys. USDA-CS Handout 436. U.S. Government printing office, Washington, DC.

Somasiri, S., S.Y. Lee, and P.M. Huang. 1971. Influence of certain pedogenic factors on potassium reserves of selected Canadian Prairie soils. *Soil Sci. Soc. Am. Proc.* 35:500–505.

Song, S.K., and P.M. Huang. 1983. Dynamics of potassium release from potassium-bearing minerals as influenced by oxalic and citric acids. *Agron. Abstr.* American Society of Agronomy, Madison, WI, p. 222.

Sparks, D.L. 1980. Chemistry of soil potassium in Atlantic Coastal Plain soils: A review. *Commun. Soil Sci. Plant Anal.* 11:435–449.

Sparks, D.L. 1985. Kinetics of ionic reactions in clay minerals and soils. *Adv. Agron.* 38:231–266.

Sparks, D.L. 1986. Kinetics of reactions in pure and in mixed systems. In: D.L. Sparks (ed.) *Soil Physical Chemistry*. CRC Press, Boca Raton, FL, pp. 83–145.

Sparks, D.L., and P.M. Huang. 1985. Physical chemistry of soil potassium. In: R.D. Munson (ed.) *Potassium in Agriculture*. American Society of Agronomy, Crop Science Society of America, and Soil Science Society of America, Madison, WI, pp. 201–276.

Sparks, D.L., and P.M. Jardine. 1981. Thermodynamics of potassium exchange in soil using a kinetics approach. *Soil Sci. Soc. Am. J.* 45:1094–1099.

Sparks, D.L., and P.M. Jardine. 1984. Comparison of kinetic equations to describe K-Ca exchange in pure and in mixed systems. *Soil Sci.* 138:115–122.

Sparks, D.L., and W.C. Liebhardt. 1982. Temperature effects on potassium exchange and selectivity in Delaware soils. *Soil Sci.* 133:10–17.

Sparks, D.L., and J.E. Rechcigl. 1982. Comparison of batch and miscible displacement techniques to describe potassium adsorption kinetics in Delaware soils. *Soil Sci. Soc. Am. J.* 46:875–877.

Sparks, D.L., D.C. Martens, and L.W. Zelazny. 1980a. Plant uptake and leaching of applied and indigenous potassium in Dothan soils. *Agron. J.* 72:551–555.

Sparks, D.L., L.W. Zelazny, and D.C. Martens. 1980b. Kinetics of potassium exchange in a Paleudult from the Coastal Plain of Virginia. *Soil Sci. Soc. Am. J.* 44:37–40.

Sparks, D.L., L.W. Zelazny, and D.C. Martens. 1980c. Kinetics of potassium desorption in soil using miscible displacement. *Soil Sci. Soc. Am. J.* 44:1205–1208.

Sposito, G. 1981. *The Thermodynamics of Soil Solutions*. Clarendon Press, Oxford.

Sposito, G. 1985. Thermodynamics of the soil solution. In: D.L. Sparks (ed.) Soil Physical Chemistry. CRC Press, Boca Raton, Fl, pp. 147–178.

Stahlberg, S. 1959. Studies on the release of bases from minerals and soils. I. The release of potassium from potassium feldspar and mica in contact with synthetic ion exchangers. *Acta. Agric. Scand* 9:361–369.

Stitcher, H. 1972. Potassium in allophane and zeolites. *Proc. Colloq. Int. Potash Inst.* 9:43–51.

Talibudeen, O., and S.K. Dey. 1968. Potassium reserves in British soils. Parts I and II. *J. Agric. Sci.* 71:95–104, 405–411.

Talibudeen, O., and K.W.T. Goulding. 1983. Apparent charge heterogeneity in kaolins in relation to their 2:1 phyllosilicate content. *Clays Clay Miner.* 31:37–42.

Talibudeen, O., and A.H. Weir. 1972. Potassium reserves in a "Harwell" series soil. *J. Soil Sci.* 23:456–474.

Talibudeen, O., J.D. Beasley, P. Leone, and N. Rajendran. 1978. Assessment of soil potassium reserves available to plant roots. *J. Soil Sci.* 29:207–218.

Tanji, K.K. 1982. Modeling of the soil nitrogen cycle. In: F.J. Stevenson (ed.) *Nitrogen in Agricultural Soils*. American Society of Agronomy, Madison, WI, pp. 721–772.

Taylor, A.W., and E.L. Gurney. 1965. Precipitation of phosphate by iron oxide and aluminum hydroxide from solutions containing calcium and potassium. *Soil Sci. Soc. Am. Proc.* 29:18–22.

Taylor, A.W., W.L. Lindsay, E.O. Huffman, and E.L. Gurney. 1953. Potassium and ammonium taranakite, amorphous aluminum phosphate and variscite as sources of phosphate for plants. *Soil Sci. Soc. Am. Proc.* 27:145–151.

Thomas, G.W. 1977. Historical developments in soil chemistry: Ion exchange. *Soil Sci. Soc. Am. J.* 41:230–238.

Udo, E.J. 1978. Thermodynamics of potassium-calcium and magnesium-calcium exchange reactions on a kaolinitic soil clay. *Soil Sci. Soc. Am. J.* 42:556–560.

van der Plas, L. 1966. *The Identification of Detrital Feldspars*. Elsevier North-Holland, New York.

van Olphen, H. 1971. Amorphous clay materials. *Science* (Washington, DC) 171:90–91.
van Reeuwijk, L.P., and J.M. de Villiers. 1968. Potassium fixation by amorphous aluminosilicate gels. *Soil Sci. Soc. Am. Proc.* 32:230–240.
Van Riemsdjik, W.H., and F.A.M. de Haan. 1981. Reactions of orthophosphate with a sandy soil at constant supersaturation. *Soil Sci. Soc. Am. J.* 45:261–266.
von Reichenbach, H.G. 1972. Factors of mica transformation. *Proc. Colloq. Int. Potash Inst.* 9:33–42.
Wada, K. 1977. Allophane and imogolite. In: J.B. Dixon and S.B. Weed (eds.) *Minerals in Soil Environments*. Soil Science Society of America, Madison, WI, pp. 603–638.
Wada, K., and Y. Harada. 1969. Effects of salt concentration and cation species on the measured cation exchange capacity of soils and clays. *Proc. Int. Clay Conf., Tokyo* 1:561–571.
Walker, G.F. 1950. Trioctahedral minerals in the soil-clays of northeast Scotland. *Mineral. Mag.* 29:72–84.
Way, J.T. 1850. On the power of soils to absorb manure. *J. Roy Agric. Soc. Engl.* 11:313–379.
Weed, S.B., C.B. Davey, and M.G. Cook. 1969. Weathering of mica by fungi. *Soil Sci. Soc. Am. Proc.* 33:702–706.
Wells, C.B., and K. Norrish. 1968. Accelerated rates of release of interlayer potassium from micas. *Trans. 9th Int. Congr. Soil Sci.* 2:683–694.
White, J.L. 1950. Transformation of illite into montmorillonite. *Soil Sci. Soc. Am. Proc.* 15:129–133.
Wiklander, L. 1949. Adsorption equilibria between ion exchangers of different nature. *Ann. Agric. College Sweden* 16:670–682.
Wilson, M.J. 1975. Chemical weathering of some primary rock-forming minerals. *Soil Sci.* 119:349–355.
Wollast, R. 1967. Kinetics of the alteration of K-feldspar in buffered solutions at low temperature. *Geochim. Cosmochim. Acta* 31:635–648.
Yuan, L.L., L.W. Zelazny, and A. Ratanaprasotporn. 1976. Potassium status of selected Paleudults in the lower Coastal Plain. *Soil Sci. Soc. Am. Proc.* 40:229–233.
Zasoski, R.J., and R.G. Burau. 1978. A technique for studying the kinetics of adsorption in suspensions. *Soil Sci. Soc. Am. J.* 42:372–374.

# Models to Assess the Susceptibility of Soils to Excessive Compaction*

S.C. Gupta and R. R. Allmaras

I. Introduction .... 65
II. Definitions .... 66
III. Modeling Laboratory Soil Compaction .... 68
A. Experimental Setups .... 68
B. Laboratory Soil Compaction Models .... 70
IV. Ease/Susceptibility of Soils to Compaction .... 77
V. Excessive Compaction .... 78
VI. Water Contents and Applied Stresses Not Conducive to Excessive Compaction .... 83
VII. Acceptable Loads Versus Drainage .... 84
VIII. Modeling Field Soil Compaction .... 86
A. Models Based on Strain Distribution .... 87
B. Models Based on Stress Distribution .... 88
IX. Application of Field Soil Compaction Models .... 94
X. Areas of Future Research .... 97
References .... 98

## I. Introduction

Compaction of soil has become a problem of worldwide concern because of (1) larger tractors, tillage, and harvesting equipment; (2) increases in intensive cropping practices; and (3) new lands being brought into cultivation, particularly in the tropics. According to McKibben (1971), between

---

*Contribution from the Department of Soil Science and the Minnesota Agricultural Experiment Station, University of Minnesota, St. Paul, MN 55108 and the Soil and Water Management Unit, USDA-ARS, St. Paul, MN, USA.

1948 and 1968 the average weight of tractors increased from 2.7 to 4.5 tons. At present, the average is 6.8 tons, with larger units weighing more than 22.4 tons.

Current research has shown that with larger, heavier field equipment and wider wheels, compaction of soil has increased at depths below the usual tilled zone of approximately 20 cm (Soehne, 1976; Voorhees *et al.*, 1978). This is of particular concern because it is difficult and costly to loosen the soil at these greater depths. Gill (1971) estimated an annual loss of $1.18 billion in the United States from soil compaction. This loss results from decreased yields and increased energy costs during field tillage.

Excessive compaction of soil reduces plant root proliferation in the soil and lowers the rate of movement of water and air through the soil (Bowen, 1981; Unger *et al.*, 1981; Wolf and Hadas, 1984). Because of root restriction, the amount of water and nutrients available to a crop are often decreased. Slower internal drainage impedes water redistribution and drain performance, prolongs the time when the soil is too wet for tillage, and decreases crop production.

Soils are often tilled when wet, causing severe compaction, yet there is no scientific method applicable to agricultural soils that determines when a soil is too wet to till. Currently, the decision must be made by the farm operator based on experience.

There is a scarcity of reliable information concerning soil compaction under field conditions that can be related either theoretically or statistically to laboratory measurements of soil compaction. If scientific models could be developed that would (1) predict the effect of applied mechanical stress on soil compaction and (2) separate beneficial from harmful compaction, simple guidelines could be developed for extension specialists and farmers. These guidelines would include (1) maximum pressures a specific soil can withstand over a range of water contents, (2) ranges of applied pressures and soil water contents that are not conducive to excessive compaction, (3) soil types and areas in a geographical region that are susceptible to excessive compaction, and (4) dominant soil and plant limiting conditions in a given area. Additionally, computer models can be used to evaluate management systems that may prevent or alleviate excessive compaction.

## II. Definitions

Before existing soil compaction models are evaluated, it is appropriate to differentiate between several terms often used in the literature to define soil compaction. Definitions described below are taken from Bradford and Gupta (1986).

*Compression.* "Compression" refers to a process that describes the decrease in soil volume (soil densification) under an externally applied load.

Applied loads can be in the form of vibration, rolling, tamping, etc. Compression of saturated soils is called consolidation, whereas compression of unsaturated soils is called compaction. During consolidation water is excluded from the voids of the soil matrix, whereas during compaction soil air is excluded from the voids of the soil matrix. Consolidation tests represent extreme conditions during which maximum densification of the soil is possible. Because of these characteristics, results of consolidation tests are used in the design of building foundations. Soils on farms are often unsaturated and so data from consolidation tests have limited application to agriculture. The property that describes the compression process is called soil compressibility.

*Compressibility*. Soil compressibility refers to the ease with which soil decreases in volume when subjected to a mechanical load. It is the slope of the straight line portion of void ratio, $e$, or bulk density $\rho$ vs. logarithm of stress (Figure 1). For saturated soils, the stress is plotted in terms of effective stress, $\sigma'$, defined as the difference between the total applied stress, $\sigma$, and the pore water pressure, $U_w$:

$$\sigma' = \sigma - U_w \quad [1]$$

Thus, for saturated compressibility, the compression index is defined as:

$$C_c = de/d \log \sigma' \quad [2]$$

For unsaturated soils, $U_w$ in equation [1] is replaced with a fraction of the pore water suction, $S$. This fraction, $\chi$, represents the normalized cross-sectional area occupied by soil water:

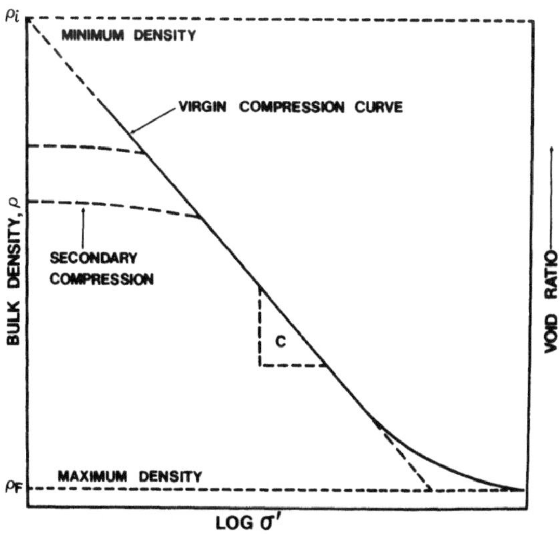


**Figure 1.** Theoretical soil compression curve.

$$\sigma' = \sigma + \chi(S) \quad [3]$$

Because of the difficulty of determining $\chi$ in equation [3], Larson *et al.* (1980), Stone and Larson (1980), and Larson and Gupta (1980) chose to plot stress in terms of the applied external stress and defined compressibility of unsaturated soil as:

$$C = d\rho/d \log \sigma \quad [4]$$

where $\rho$ is the bulk density.

*Compactibility*. Compactibility is the maximum density to which a soil can be packed by a given amount of energy. The American Society for Testing and Materials (ASTM) (1985) gives a standard method for determining soil compactibility. In this procedure, often called the Proctor test, the soil is compacted in a metal mold by a set number of impacts from a free-falling hammer. The soil density obtained with this procedure at a series of soil water contents is used to define compactibility (maximum density) for a given energy level.

*Compaction*. Compaction has been used for both laboratory and field soil compression characterizations. In the laboratory, soil compaction refers to the compression of small homogeneous soil samples, whereas in the field it refers to the simultaneous densification of several soil horizons in the profile. Compressibility generally describes soil compaction in the laboratory. However, no single property is available to describe soil compaction in the field.

Soil compaction has often been used or understood to have bad connotations; however, not all soil compaction is bad. There are times when seedbed compaction is needed to improve soil–seed contact. Thus, it is necessary to differentiate good or acceptable compaction from harmful or excessive compaction.

## III. Modeling Laboratory Soil Compaction

### A. Experimental Setups

Soil compaction tests in the laboratory are conducted under either a triaxial or a uniaxial compression setup. Figure 2 shows a cross section of a typical triaxial cell and the forces acting on a soil sample in a triaxial test. In this test, samples held by a membrane are subjected to applied stresses in the vertical ($\sigma_1$) and horizontal ($\sigma_3$) directions. After each application of load, the samples are allowed to equilibrate and measurements are made of the axial strain (length or density of the samples). Various combinations of $\sigma_1$ and $\sigma_3$ are used to measure the changes in axial strain.

Figure 3 shows the experimental setup of a uniaxial compression test. In this test, remolded or undisturbed samples are subjected to applied loads in the vertical direction only. The cells described in the uniaxial

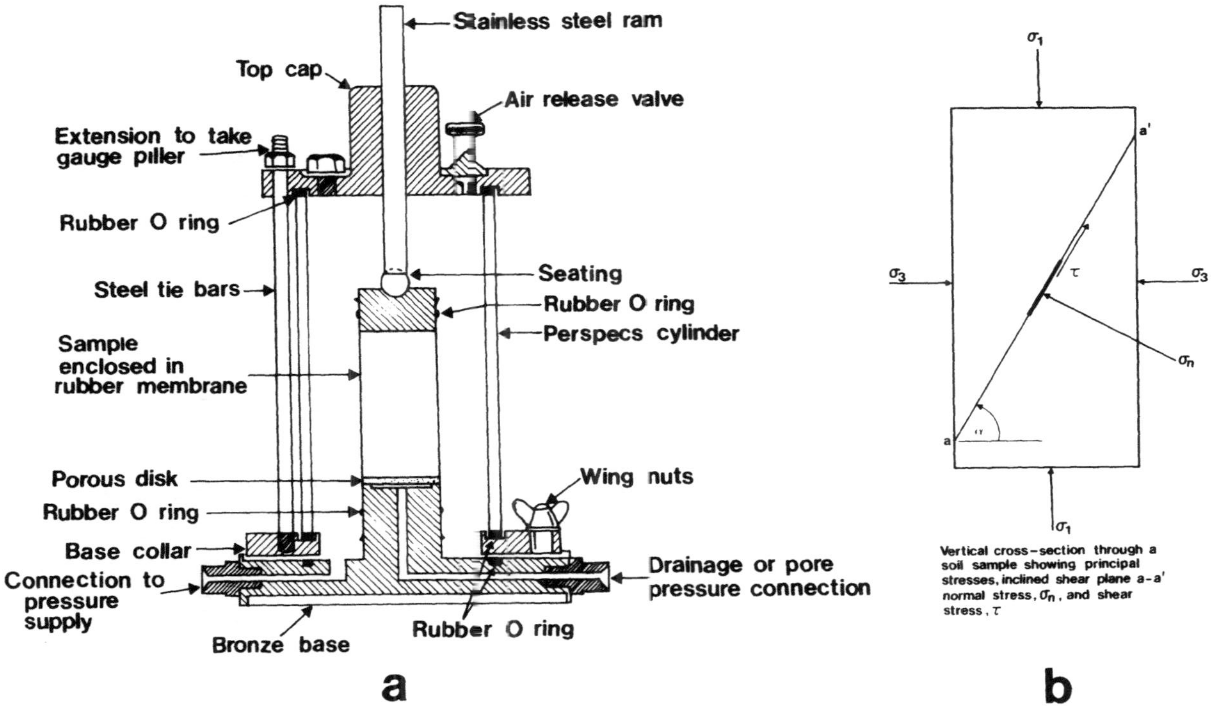


**Figure 2.** (a) Cross section of a typical triax al cell. From Bishop and Henkel (1962). (b) Stress acting on a soil sample in a triaxial compression test. From Jumikis (1964).

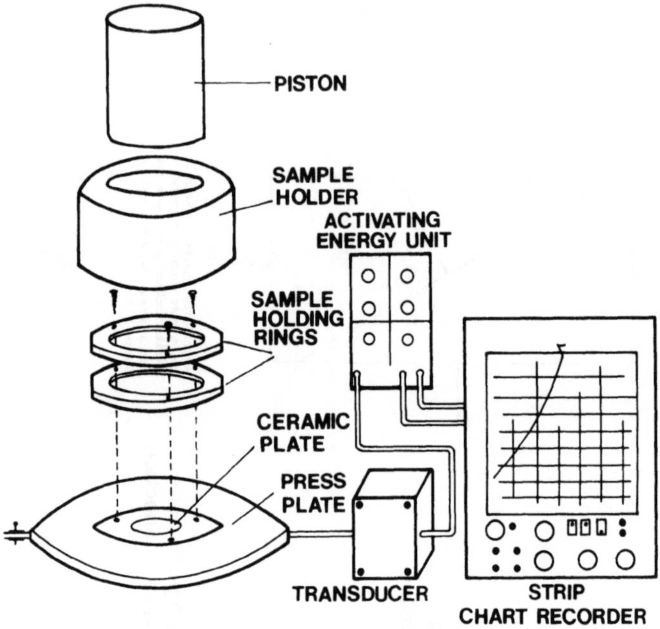


**Figure 3.** Apparatus used for uniaxial compression determination. From Larson *et al.* (1980).

compression test are called odeometers or consolidometers. Changes in the density of soil samples are measured after equilibration with the applied load or after set times or continuously. The uniaxial compression test is rather quick but has the disadvantage of an unknown $\sigma_3$ acting on the sample.

### B. Laboratory Soil Compaction Models

Laboratory soil compaction models are generally various strategies for representing stress–strain relations in a small sample. An example of the stress–strain relationship obtained in the triaxial setup is shown in Figure 4. At a given $\sigma_1$, axial strain increases with a decrease in $\sigma_3$. Because the triaxial tests are time consuming and expensive, Perumpral *et al.* (1971) suggested a normalizing procedure to describe stress–strain relationships at several $\sigma_3$ values derived from one or two measured curves. The normalizing procedure involved calculating normalized octahedral strains and stresses from measured values of $\sigma_1$, $\sigma_2$, $\sigma_3$, $\varepsilon_1$, $\varepsilon_2$, and $\varepsilon_3$ using equations [5] and [6]:

$$\gamma_o = 2/3\,[(\varepsilon_1 - \varepsilon_2)^2 + (\varepsilon_2 - \varepsilon_3)^2 + (\varepsilon_3 - \varepsilon_1)^2]^{1/2} \qquad [5]$$

$$\tau_o = 1/3\,[(\sigma_1 - \sigma_2)^2 + (\sigma_2 - \sigma_3)^2 + (\sigma_3 - \sigma_1)^2]^{1/2} \qquad [6]$$

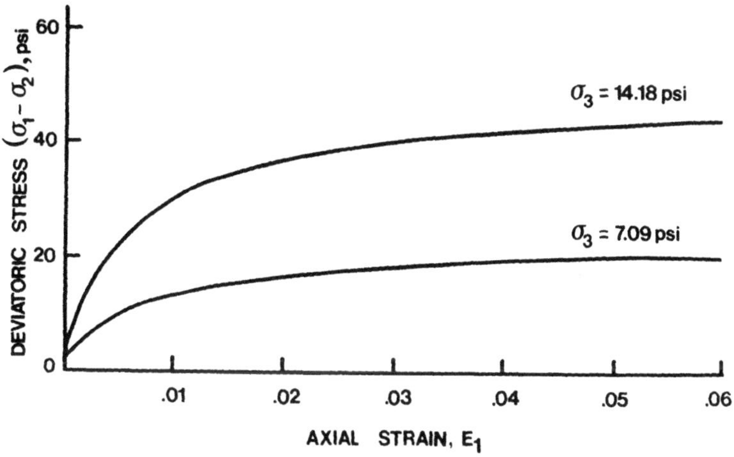


**Figure 4.** Stress-strain relationship from triaxial tests. From Perumpral *et al.* (1971).

Subscripts 1, 2, and 3 on stress ($\sigma$) and strain ($\varepsilon$) symbols correspond to *x*, *y*, and *z* coordinates. Figure 5 shows the plot of normalized shear stress vs. normalized shear strain for data shown in Figure 4.

Most efforts in modeling laboratory compression tests have been in developing mathematical formulations to describe the curves in Figures 4 and 5. Coleman and Perumpral (1974) approximated the curve in Figure 5 in two parts; a broken line as the linear relationship:

$$\gamma_0 = 0.004641\ (\tau_0)_n \qquad \text{for} \qquad (\tau_0)_n < (\tau_0)_y \tag{7}$$

and a solid line as the third-degree polynominal:

$$\gamma_o = 0.01409(\tau_o)_n - 0.04169(\tau_o)_n^2 + 0.04558(\tau_o)_n^3 \quad \text{for } (\tau_o)_n \geq (\tau_o)_y \tag{8}$$

where $(\tau_0)_n = \tau_0/(\sigma_0)_i$ is the normalized octahedral shear stress, $(\tau_0)_y$ is the initial octahedral normal stress (confining stress).

In developing a model to describe laboratory compaction, Bailey *et al.* (1985) suggested that the model should at least fit the following two boundary conditions. At high applied compressive stresses ($\sigma_a$), the slope of the curve should be constant:

$$d\bar{\varepsilon}_v/d\sigma_a = \text{constant} \tag{9}$$

where $\bar{\varepsilon}_v = \ln (V/V_0)$ is the natural volumetric strain, $V$ is the specific volume at a given stress and $V_0$ is the initial specific volume in an unstressed state. In other words, as a soil approaches its maximum bulk density, additional compressive stress may cause an apparent elastic behavior so that a change in compressive stress causes a proportional change in volumetric strain. The second boundary condition implies no volumetric strain when applied stress is zero.

A model that describes the shape of the hydrostatic compaction curve

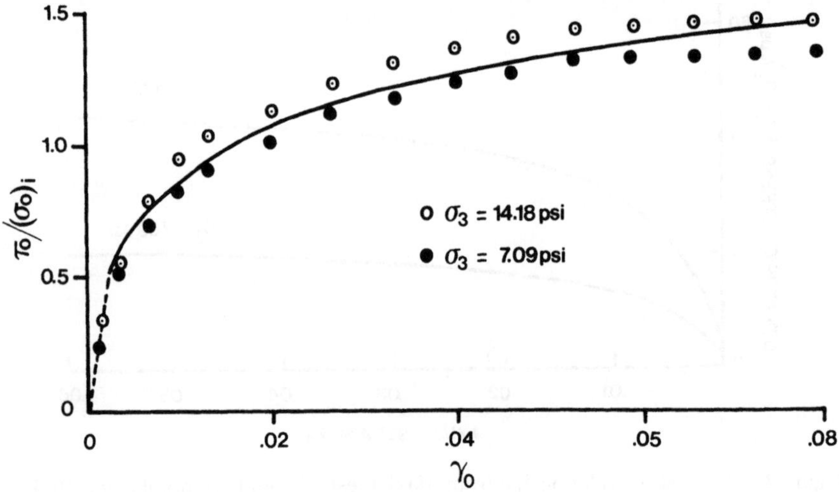


**Figure 5.** Normalized stress–strain relationship used for finite element analysis. From Coleman and Perumpral (1974).

and also fits the two boundary conditions as defined above is the following (Bailey *et al.*, 1985):

$$\bar{\varepsilon}_v = (A + B\,\sigma_h)[1 - \exp(-C\sigma_h)] \qquad [10]$$

where $\sigma_h = \sigma_1 = \sigma_2 = \sigma_3$ = hydrostatic confining stress. Assuming a constant mass of soil, equation [10] can also be expressed in terms of $\rho$:

$$\ln \rho = \ln \rho_0 - (A + B\sigma_h)[1 - \exp(-C\sigma_h)] \qquad [11]$$

where $\rho_0$ is the bulk density at zero stress.

Using the data obtained from hydrostatic triaxial tests, Bailey *et al.* (1985) showed that equation [11] is a better model to describe laboratory compaction curves (Figure 6) than an earlier simple model (Bailey and Vanden Berg, 1967):

$$1/\rho = m \log \sigma_a + b \qquad [12]$$

where $m$ is the compression index and $b$ is the intercept, which is equal to $1/\sigma$ at a stress value of 1.0.

The statistical fit of equation [11] to triaxial data obtained on Lloyd clay (Typic Hapludult) at various soil water contents showed a dependence of $\rho_0$ and coefficients $A$, $B$, and $C$ on water content (Figure 7).

Dexter and Tanner (1973) proposed equation [13] to describe hydrostatic compression in a triaxial setup (Figure 8). They suggested the use of particle packing density, $D$, instead of bulk density as measured during the triaxial tests:

$$D = D_0 + B \exp(-K\sigma_a) + C \exp(-L\sigma_a) \qquad [13]$$

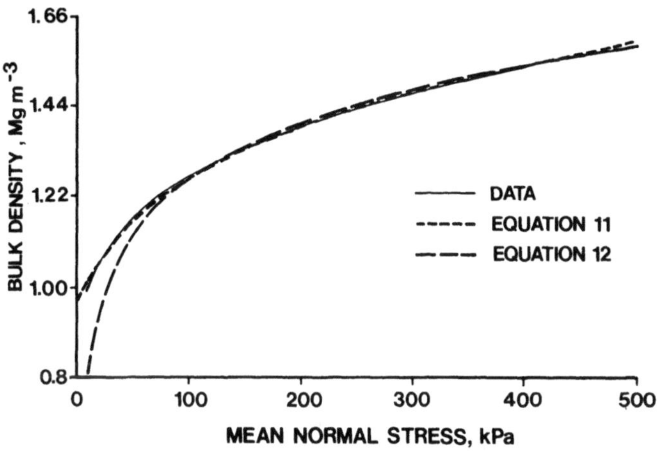


**Figure 6.** Comparison of triaxial compression data (solid line) for Lloyd clay (Typic Hapludult) at 18.4% moisture content with equation [11] (short dash) and equation [12] (long dash). From Bailey *et al.* (1985).

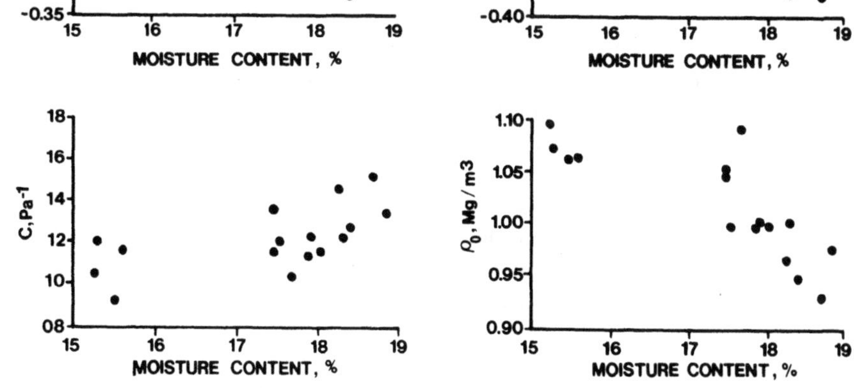


**Figure 7.** Influence of moisture content on coefficients A, B, and C, and $\rho_0$ in equation [11] for Lloyd clay (Typic Hapludult).

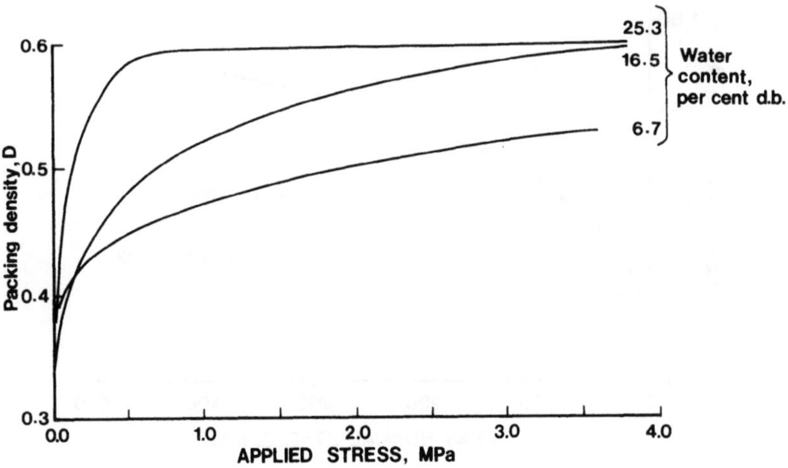


**Figure 8.** Increase in particle packing density with applied pressure in a sandy clay loam soil at three water contents. From Dexter and Tanner (1973).

where $D_0$ is the maximum limiting packing density, $(D_0 + B + C)$ is the initial packing density, and $K$ and $L$ are measures of how rapidly the maximum packing density is attained by increasing $\sigma_a$. Particle packing density is defined as:

$$D = (\rho/2.66) \times (100 - OC)/(100 + W) \qquad [14]$$

where 2.66 Mg m$^{-3}$ (the density of quartz) is the average density for individual soil particles, $OC$ is the percentage organic carbon, and $W$ is water content by weight. The particle packing density, $D$, is the proportion of the total soil volume that is occupied by mineral particles. The packing density rises asymptotically to a maximum limiting value and the rate of rise is greater at higher water contents.

By fitting equation [13] to laboratory compaction curves of five soils at several water contents, Dexter and Tanner (1973) inferred that $B$ and $K$ corresponded to interactions between individual soil particles, whereas parameters $C$ and $L$ corresponded to the change in soil crumb structure. The value of $L/K$ was between 10 and 40, indicating the destruction of crumbs at much lower stresses than required for a significant rearrangement of individual soil particles.

Larson *et al.* (1980) developed a model similar to that of Bailey and Vanden Berg (1967) to describe the straight line portion of the virgin compression curve (Figure 1):

$$\rho = \rho_K + C \log (\sigma_a/\sigma_K) \qquad [15]$$

where $\rho_K$ is the density at a known stress, $\sigma_K$, and $C$ is the slope of the line (equation [4]) in Figure 1, which also was defined as the compression index. Larson *et al.* (1980) suggested that the virgin compression curve was the only portion of the compression curve that was important to the prediction of compaction of agricultural soils. The reasons were that (1) along the straight line portion of the compression curve the soil sample does not return to its initial state when stress is removed after application, (2) secondary compression reflects the previous history (stresses and aggregate size distribution) of the sample, and (3) there is little change in compression of the sample because of applied stress during secondary compression. Based on the slope of the virgin compression curve, Larson *et al.* (1980) grouped agricultural soils into four compression categories (Figure 9). Curve A represents the amorphous materials, such as allophanes; curves B and C represent medium- and fine-textured soils from the tropics and temperate regions of the world, respectively; and curve D represents sandy soils. Differences in the slopes of the virgin compression curves of B and C result from variations in the amount of organic matter present and the type of clay minerals. Temperate region soils are high in swelling (2 : 1) clay minerals and organic matter, whereas tropical region soils contain small amounts of organic matter and are dominated by nonswelling (1 : 1) clay minerals.

Because triaxial tests are expensive and time consuming, Larson *et al.* (1980) used a uniaxial compression setup to obtain laboratory compaction

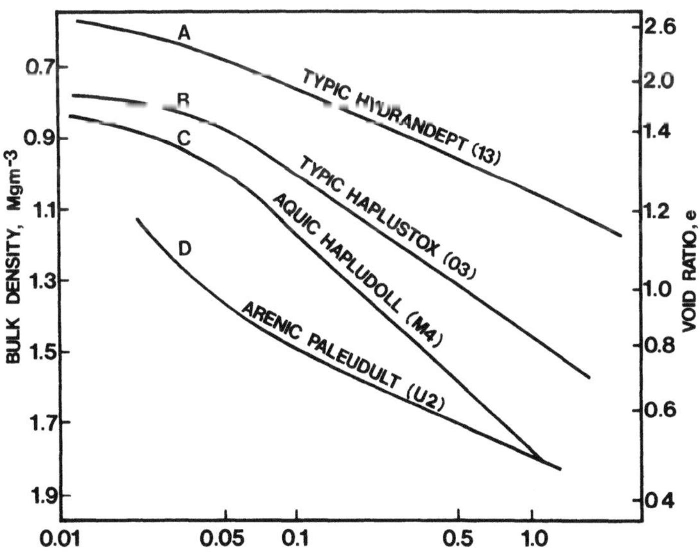


**Figure 9.** Soil compression curves for a Typic Hydrandept, Typic Haplustox, Aquic Hapludoll, and Arenic Paleudult at approximate pore water potentials of −30 kPa. From Larson *et al.* (1980).

curves. Therefore, the data reported by Larson *et al.* (1980), Larson and Gupta (1980), and Gupta *et al.* (1985a, b) refer to changes in linear strain caused by applied load as compared to volumetric strain measured in the triaxial setup. Using the principles of "critical state soil mechanics" and the triaxial data reported in the literature on agricultural soils, Koolen and Kuipers (1983) suggested that bulk density varied mainly according to $\sigma_1$ and only a little with $\sigma_3$. Thus, triaxial apparatus is not required and compaction of loose soil in a rigid cylinder by a piston (uniaxial compression test) is sufficient. Larson *et al.* (1980) modified the conventional uniaxial compression setup to monitor the changes in pore-water pressure during compression. These data were used to define a critical stress above which aggregates sheared and their identity was destroyed (Larson and Gupta, 1980).

Uniaxial compaction curves of Larson *et al.* (1980) showed that the virgin compression curve shifted to the right with a decrease in water content (initial pore pressure of −5 to −60 kPa) of the soil (Figure 10). Additionally, bulk density at $\sigma_K$ = 100 kPa was linearly related to the degree of water saturation corresponding to an initial pore water pressure of −5 to −60 kPa. Thus, Larson *et al.* (1980) combined the equation of the line relating bulk density at $\sigma_K$ = 100 kPa to degree of saturation with equation [15] to describe the effects of water content on laboratory compaction curves:

$$\rho = [\rho_{K1} + \Delta_T(S_1 - S_K)] + C \log (\sigma_a/\sigma_K) \qquad [16]$$

where $\rho_{K1}$ is the bulk density at a reference degree of water saturation, $S_K$, and a reference applied stress, $\sigma_K$; $\Delta_T$ is the slope of $\rho_K$ vs. degree of water saturation, and $S_1$ is the desired degree of water saturation. Based on the data from 54 and 87 samples of soil horizons from nine orders, Gupta and Larson (1982) and Gupta *et al.* (1985a), respectively, provided

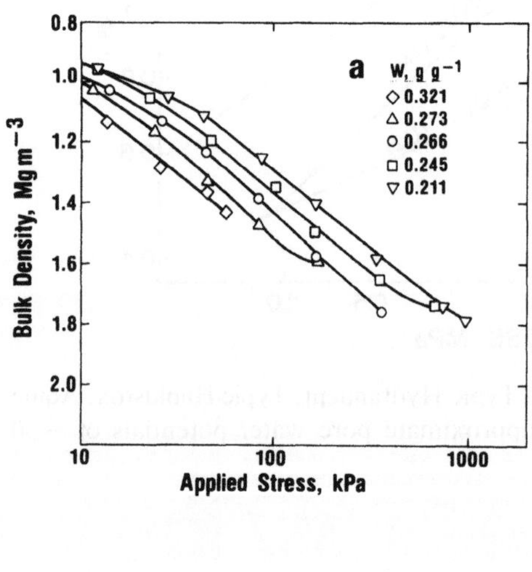


**Figure 10.** Bulk density of Waukegan silt loam (Typic Hapludoll) as influenced by applied stress at five water contents.

relationships of various parameters in equation [16] to particle size distribution.

In Figure 10, virgin compression curves are nearly parallel to each other at various water contents. Thus, Larson *et al.* (1980) calculated an average value for the slope of the virgin compression line that applied over an initial pore-water pressure of $-5$ to $-60$ kPa. Standard deviation of the average compression index was higher for fine-textured soils as compared to sand and loamy sands.

## IV. Ease/Susceptibility of Soils to Compaction

Ease/susceptibility of soil to compaction implies the rate at which soil compresses with applied load at a given degree of saturation. In other words, ease/susceptibility of soils to compaction is a property of the soil and is analagous to the compression index. Figures 11 and 12 show the relationship of the compression index to clay content for soils with expanding and nonexpanding clay types, respectively. Considering the scatter of data in Figures 11 and 12, the compression index does not show much difference between temperate and tropical soils of the same texture. This may be partially (1) because of the finite thickness of the soil samples in the laboratory compression tests, (2) because of the destruction of soil structure during initial preparation (grinding) of the samples, and (3) because applied stresses along the virgin compression curve are much higher than the swelling pressure exerted by various clays. The compression index increases with increasing clay content up to 33% clay and then levels off. Higher values of the compression index also signify higher compressibility or greater compaction of a soil. Thus, Figures 11 and 12 indicate the ease/susceptibility of soils to compaction by textural class. Figures 11 and 12

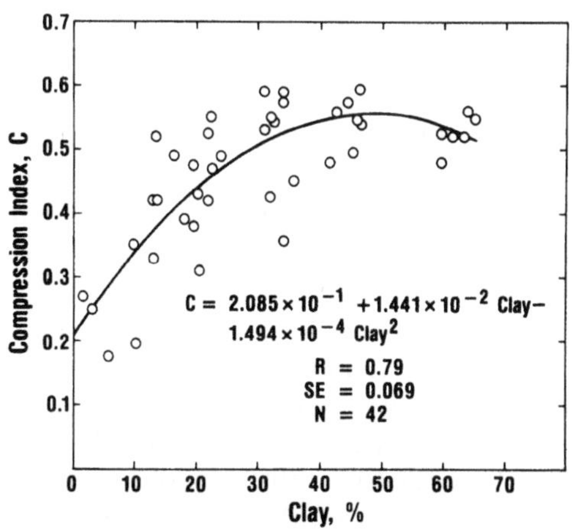


**Figure 11.** Relationships between compression index and percentage clay for Mollisols, Spodosols, Entisols, Inceptisols, and Vertisols. From Gupta *et al.* (1985a).

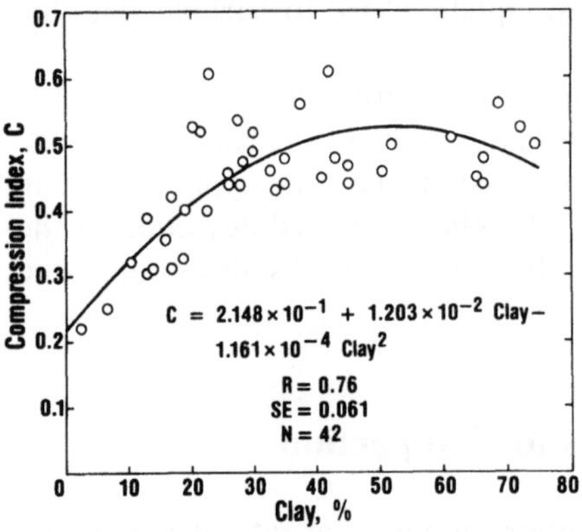


**Figure 12.** Relationship between compression index and percentage clay for Alfisols, Ultisols, and Oxisols. From Gupta *et al.* (1985a).

show that clay soils are most susceptible while sandy soils are least susceptible to compaction.

However, soils with higher compression indexes are not necessarily unsatisfactory for root and plant growth. Soils with larger compression indexes simply exhibit larger changes in the air–water–soil matrix relationship for an increment of applied load. It is possible that these soils have a less than limiting air–water–soil matrix relationship (lower $\rho$), even after a load has been applied. Hence, the ease/susceptibility of soils to compaction does not give enough information to describe degradation of soils as a medium for plant growth. What is needed is a comparison of air–water–soil matrix relationships at various applied loads to the limiting values for root and shoot growth. This comparison will not only define the relative ease/susceptibility of soils to excessive compaction but also delineate the range of stresses and soil water contents that will not be conducive to excessive compaction.

## V. Excessive Compaction

Gupta and Larson (1982) suggested three variables to define soil and plant limiting conditions arising from soil compaction. These include (1) the air-filled porosity critical for gaseous diffusion, (2) stresses critical for shearing soil aggregates, and (3) soil resistance critical for root growth. However, the literature is not only limited but often confusing about the limits of these variables in defining excessive compaction.

Gupta and Larson (1982) assumed 10% air-filled porosity as a critical value below which gaseous exchange with the atmosphere may restrict biological activities. The limit of 10% air-filled porosity is based on Figure 13, where relative diffusivity of oxygen is nearly zero. Grable and Siemer (1968) also showed that at 12% air-filled porosity, the rate of corn root

**Figure 13.** Influence of air porosity on relative diffusivity of oxygen in silty clay loam at different bulk densities and aggregate sizes. From Grable and Siemer (1968).

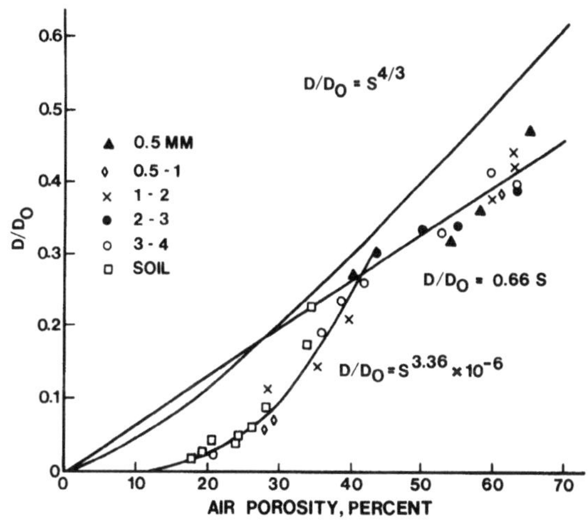


elongation is nearly zero. This limiting air-filled porosity, however, depends upon the type of plant and the level of microbial activity of the soil.

Critical stresses for shearing soil aggregates are based on the hypothesis that pore-water pressure changes when the majority of soil aggregates shear during compaction (Larson and Gupta, 1980). Figure 14 is an example of the changes in the normalized pore water pressure, $U_n$, as a function of normalized applied stress $\sigma_n$ during uniaxial compression. The stress value corresponding to the peak pore pressure is defined as the stress at which soil aggregates shear. Larson and Gupta (1980) explained that during the initial application of an applied load, the number of contacts of each aggregate increases, that is, each aggregate comes into contact with more of the surrounding aggregates. Each new contact produces a new meniscus, which acts to pull water from within the aggregates and thereby reduces the pore-water pressure. As additional mechanical stress is added, the new contact area increases more slowly (little change in pore-water pressure occurs) until pores between the aggregates become small enough for the menisci to coalesce; i.e., the pores fill with water, and then the pore-water pressure increases. As the pore-water pressure approaches zero from the negative, aggregate strength decreases rapidly, whereupon further application of applied load shears the aggregates and destroys their identity.

Gupta *et al.* (1985a) provided a relationship to estimate $\sigma_n$ from clay content ($< 2\mu$):

$$\sigma_n = 0.5806 + 0.8653 \times 10^{-2}\,\text{clay} - 0.8752 \times 10^{-4}\,\text{clay}^2 \quad [17]$$

$$\sigma_n = \log \sigma_C / \log \sigma_S \quad [18]$$

where $\sigma_C$ is the critical stress above which soil aggregates shear and $\sigma_S$ is the applied load when soil is saturated at a given water content.

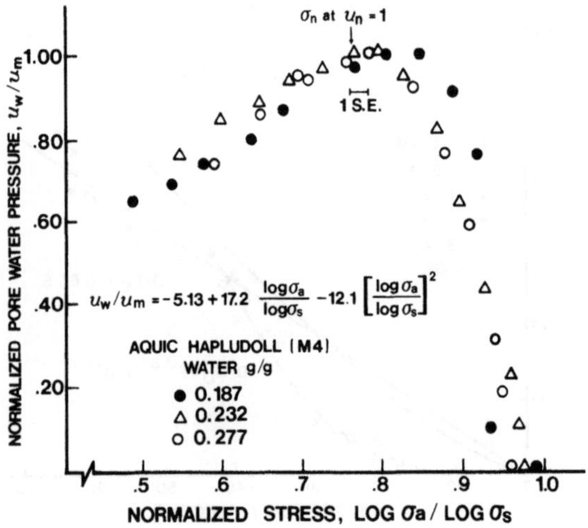


**Figure 14.** Normalized pore water pressure, $U_w/U_m$, vs. normalized stress, log $\sigma_a$/log $\sigma_s$, for an Aquic Hapludoll at three water contents. From Larson and Gupta (1980).

A penetrometer resistance of 2.0 MPa was assumed critical for root penetration (Figure 15). This is based on the relationship of cotton root penetration to penetrometer resistance (H.M. Taylor *et al.*, 1966). Voorhees *et al.* (1975) obtained a similar value for pea (*Pisum sativum* L.) seedlings in a fine-textured soil. However, Gerard *et al.* (1982) showed that penetrometer resistance above which root growth is negligible varies with the clay content (Figure 16). Additionally, critical penetrometer resistance will depend upon the type of root system and the type of penetrometer. Other disadvantages of using penetrometer resistance as an index of excessive soil compaction is an insufficient data base for penetrometer resistance as a function of soil water content and bulk density. Russell and Goss (1974) presented a relationship of root elongation to applied pressure on roots of barley (*Hordeum vulgare* L.) plants grown 6 days in

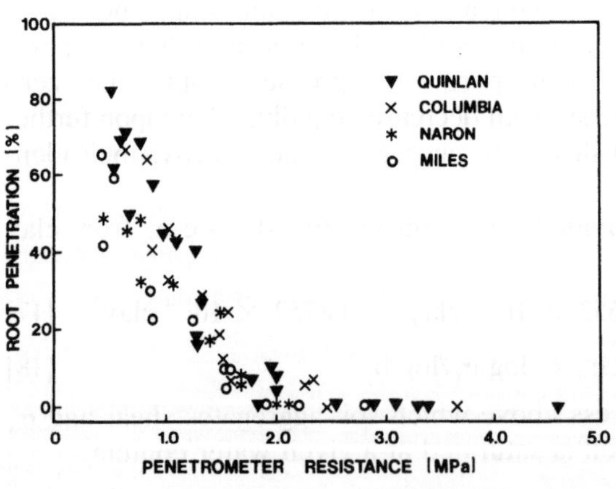


**Figure 15.** The effect of penetrometer resistance on the percentage of cotton (*Gossypium hirsutum* L.) tap roots that penetrated through 0.025 m thick cores of four soils. From Taylor *et al.* (1966).

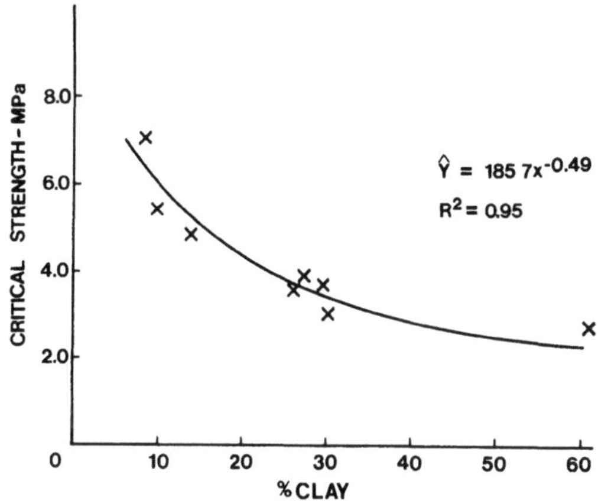


**Figure 16.** Relationship between critical soil strength (which stopped root elongation) and percentage clay. From Gerard *et al.* (1982).

beds of ballontini (Figure 17). The root elongation rate of barley was reduced by 50 and 80% at applied stresses of 20 and 50 kPa, respectively. Relationships such as that in Figure 17 can be used directly to compute applied loads large enough to limit root growth for maximum yields. Data such as those in Figure 17 also eliminate the need for instrument-dependent measurements, such as penetrometer resistance.

Other variables that can define excessive compaction are the hydraulic properties of soil. Excessive soil compaction impedes the flow of water and thus reduces the effectiveness of drains. This may be particularly important in fine-textured soils where drains are installed to improve the

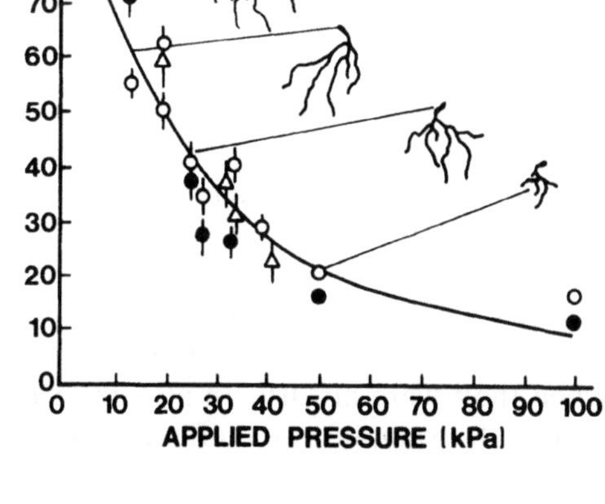


**Figure 17.** Effect of applied pressure on the rate of elongation of seminal axes of barley plants grown for 6 days in beds of ballontini. Open circles; pore diameter 157 μm; closed circles; pore diameter 69 μm; triangles; pore diameter 16 μm. From Russell and Goss (1974).

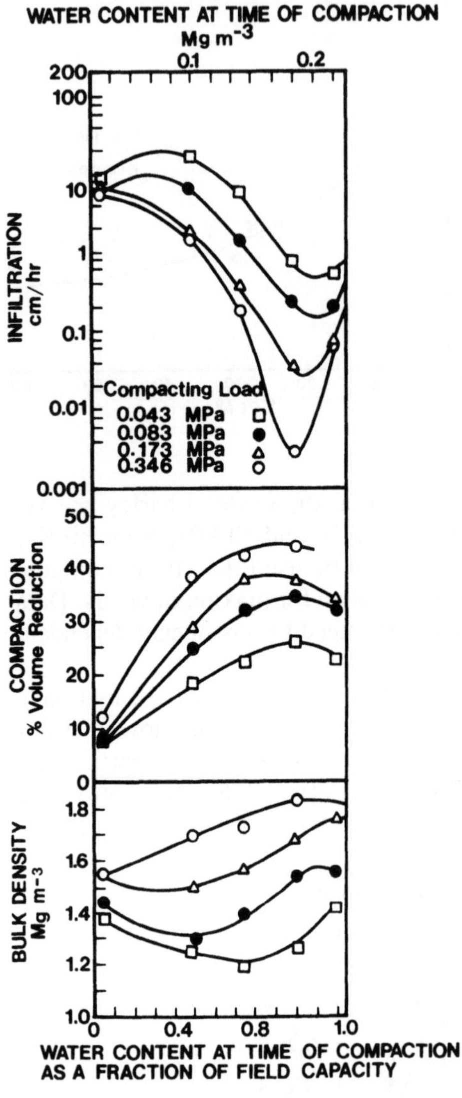


**Figure 18.** Effects of compacting loads and water content at time of compaction on infiltration rate, compaction, and bulk density of a sandy loam soil. From Akram and Kemper (1979).

inherent poor drainage characteristics. Currently, there is a limited data base on the changes in soil hydraulic characteristics as influenced by applied loads. Akram and Kemper (1979) showed that laboratory infiltration rates on five soils were a function of compacting pressure and water content at the time of compaction. Compacting loads of 0.346 MPa at field capacity on sandy loams and finer textured soils reduced infiltration rates to less than 0.1% of values obtained after these soils had been compacted when they were air dry. In a loamy sand this reduction was about 1%. Figure 18 exemplifies the effects of compacting loads and water content at the time of compaction on infiltration rate, compaction, and bulk density of a sandy loam soil (Akram and Kemper, 1979).

Critical infiltration that may define excessive compaction depends upon the rainfall intensities of the area. At a minimum, the critical infiltration rate for a given load and soil water content should be equal to the rainfall intensity. This, however, becomes complicated when the variations in rainfall intensities from year to year and the changes in hydraulic properties with soil depth are considered.

## VI. Water Contents and Applied Stresses Not Conducive to Excessive Compaction

Laboratory compaction models provide means to estimate water contents and applied stresses below which a soil is not susceptible to excessive compaction. The procedure involves superimposing critical values of variables defining excessive compaction on the air–water–soil matrix at various applied loads. An example of this procedure is shown in Figure 19, where plots of porosity vs. water content at various applied loads were calculated from the laboratory compaction model (equation [16]). Three criteria of excessive compaction (Gupta and Larson, 1982) superimposed on Figure 19 delineate the range of water contents and applied loads (the shaded area) that do not cause excessive compaction. Figure 19 could be

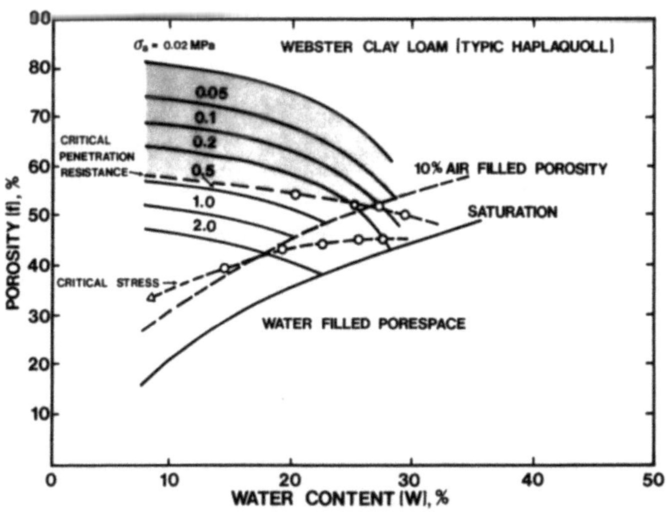


**Figure 19.** Porosity vs. water content diagram of Webster clay (Typic Haplaquoll) loam with superimposed criteria of (1) 10% air-filled porosity, (2) stress critical for shearing aggregates, and (3) critical penetration resistance.

further improved upon to delineate the range of water contents and loads that would not critically reduce the drainage performance of soils. However, a large data base of infiltration at various applied loads and water contents at the time of compaction is needed to define infiltration rates that are critical for a given rainfall intensity.

## VII. Acceptable Loads Versus Drainage

Using the laboratory compaction model, Amir *et al.* (1976) evaluated the relative importance of water content and applied stress for producing soil compaction. Between the water contents of 0.4 and 0.9 of saturation, Amir *et al.* (1976) suggested that porosity for the field soil $(N)$ can be expressed by the equation:

$$N = A_n - B_n \ln (\sigma_r + \sigma_a) - C_n \ln \Theta \qquad [19]$$

where $\sigma_r$ is the residual pressure, $\Theta$ is the volumetric water content, and $A_n$, $B_n$, and $C_n$ are constants. For virgin soils (soils without precompaction), $\sigma_r = 0$ in equation [19]:

$$N = A_n - B_n \ln (\sigma_a) - C_n \ln \Theta \qquad [20]$$

Figure 20 shows a relationship of porosity and water content for virgin and field compacted Yolo silt loam (Typic Xerosthents). Amir *et al.* (1976) defined $\sigma_r$ in equation [19] as the pressure which, when added to the applied pressure, results in residual compaction. According to these authors, $\sigma_r$ has two properties: (1) it is constant, no matter at what applied pressure it is calculated, as long as the soil water content is the same; and (2) its influence on the actual measured porosity decreases with an increase in $\sigma_a$.

Equation [20] can be rewritten to express the influence of load and water content on change in porosity of the virgin compression curve:

$$\Delta N = N_1 - N_2 = B_n \ln (\sigma_{a2}/\sigma_{a1}) + C_n \ln (\theta_2/\theta_1) \qquad [21]$$

$$\sigma_{a2}/\sigma_{a1} = [\exp (\Delta N/B_n)]/(\theta_2/\theta_1)^{(C_n/B_n)} \qquad [22]$$

Equation [22] could be used to determine when to undertake field operations involving traffic. Assuming a certain degree of compaction expressed as a change in porosity, $\Delta N$, one can operate in the field with machine stress of $\sigma_{a1}$ when soil water content is $\Theta_1$, but if one waits until the soil water content decreases to $\Theta_2$, heavier machinery that exerts a stress of $\sigma_{a2}$ can be used. For example, for a soil having compression characteristics $B_n = 4$ and $C_n = 12$, one could use machinery exerting 63% more ground stress ($P_2 = 1.63P_1$) if the water content is lowered by 15% ($\Theta_2 = 0.85\Theta_1$). Similarly one can double the ground stress ($P_2 = 2P_1$) if the water content is lowered by 20% ($\Theta_2 = 0.8\Theta_1$).

Decreases in soil water content result primarily from evapotranspiration

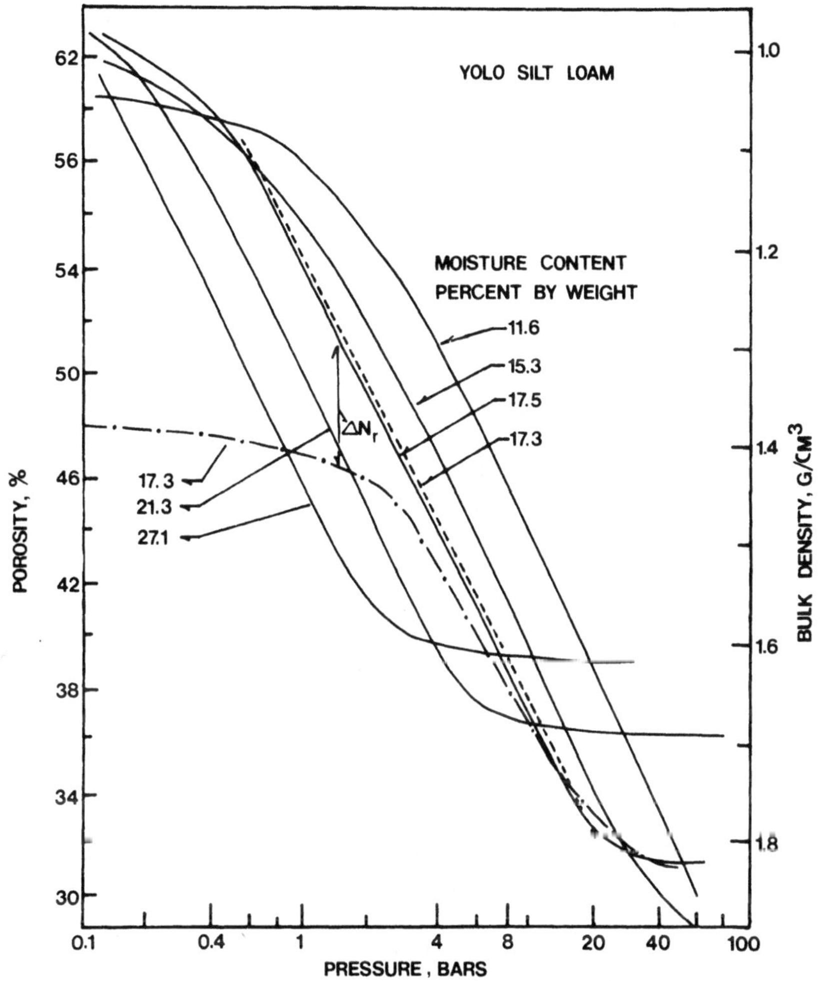


**Figure 20.** Relationship between porosity and compaction pressure for several soil moisture contents for precompacted and virgin Typic Xerorthent. From Amir *et al.* (1976).

and drainage. If natural drainage and evapotranspiration are not rapid enough, drains are usually installed to increase the rate of soil water movement. Amir *et al.* (1976) incorporated these ideas into their laboratory soil compaction model to evaluate the advantages of turnaround time arising from increased drainage by artificial means.

Hillel (1971) and Baver *et al.* (1972) presented an empirical relationship of decrease in water content in the initially wetted zones:

$$\Theta = R(t + Q)^{-S} \qquad [23]$$

where $R$, $Q$, and $S$ are constant and depend on soil drainage characteristics and $t$ is time. For $t$ greater than 1 day, $Q$ can be neglected. Thus, equation [23] can be written:

$$\Theta = Rt^{-S} \tag{24}$$

or:

$$\ln \Theta = \ln R - S \ln t \tag{25}$$

Substituting equation [25] into equation [22] and rearranging the terms results in:

$$(\sigma_{a2}/\sigma_{a1}) \exp(-\Delta N/B_n) = (t_2/t_1)^{-SC_n/B_n} \tag{26}$$

Amir *et al.* (1976) gave a graphical presentation of equation [26] (Figure 21) for a soil with laboratory compaction characteristics $C_n/B_n = 2.5$. Figure 21 shows that at 5 days after saturation, the applied stresses can be increased by 50 and 150% for $S_2/S_1 = 1$ and $S_2/S_1 = 2$, respectively, without any change in the level of compaction. Additionally, this soil can withstand the same stress ($P_2/P_1 = 1.7$) in 3.2 days as compared to 10 days if drainage characteristics are improved by a factor of 2 ($S_2/S_1 = 2$). Thus, improving soil drainage shortens the turnaround time for taking machinery on the field without changing the compaction level. Of course, the decision to install drainage would depend upon the economics of increased yield through early turnaround time as compared to the expense of installing drains.

# VIII. Modeling Field Soil Compaction

In Figure 19 a procedure is demonstrated for delineating the range of stresses and soil water contents that are not conducive to excessive compaction. However, these data apply to thin sections (small samples), where

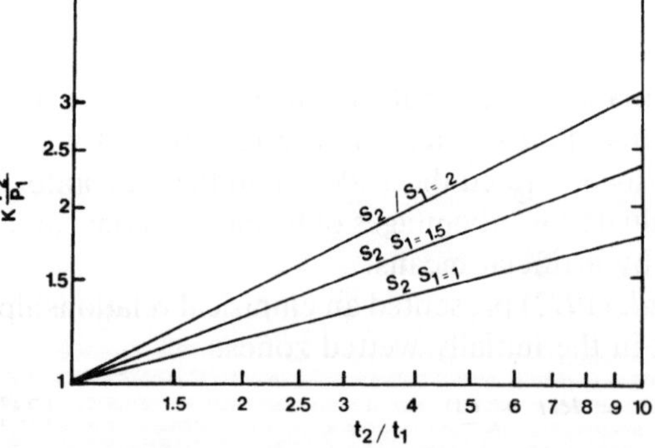


**Figure 21.** Relationship between applied stress ratio and time ratio for several drainage coefficient ratios $S_2/S_1$ (for $C_n/B_n = 2.5$) giving equal compaction. From Amir *et al.* (1976).

it is assumed that applied stress is uniformly distributed throughout the length of the sample. In the field, stresses applied at the soil surface dissipate with soil depth. Furthermore, because of stratification in field soils, changes in volume and lineal strain at various depths in the soil profile are not the same. Thus, a model is needed to describe the strain and pressure distribution profiles after the application of stresses at the soil surface.

## A. Models Based on Strain Distribution

Perumpral *et al.* (1971) and Coleman and Perumpral (1974) used the concepts of the potential energy in a finite element technique to predict, first, soil deformation and then indirectly the stress distribution under an applied load. Briefly, this method of analysis divides a soil continuum into a finite number of discrete elements called the finite elements (Figure 22). These elements are connected together at a finite number of points called nodal points. A displacement function to approximate the exact displacement field across the element is chosen. The displacement function chosen should satisfy the compatibilty conditions at all times. Material properties

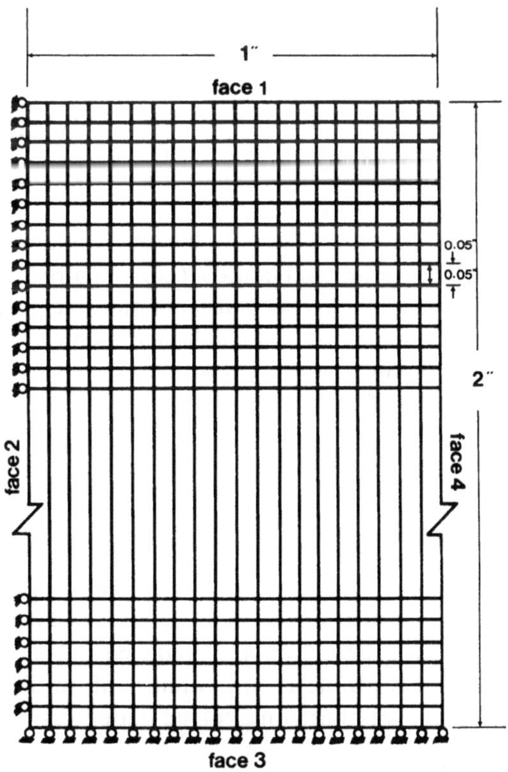


**Figure 22.** Idealized system for finite element analysis. Presence of roller indicates boundary restraints in the movement of these elements.

are ascertained from laboratory compaction tests (equations [7] and [8]) and are specified in terms of shear or deformation moduli, $G$:

$$G = d\tau_0/d\,\gamma_0 \tag{27}$$

Hooke's law requires that deformation moduli be invariant for linearly elastic materials. However, they are dependent on the state of stress for nonlinearly elastic materials. The element stiffness is computed from the known element elasticity ($E$) and the potential energy ($\pi_p$) of the element:

$$E = 2G(1 + m) \tag{28}$$

where $m$ is the Poisson ratio. The potential energy of the element is defined as the sum of the internal strain energy, $u$, and the potential of the external load, $\omega_p$:

$$\pi_p = u + \omega_p \tag{29}$$

The potential of an external load signifies the capacity of a load $P$ to perform work through a deformation $\varepsilon$ of the column. To simulate deformation of the complete system of finite elements, equations are added describing each element. The assembling process is constrained by the law of compatibility or continuity. It requires that the body remain continuous; that is, the neighboring points must remain next to each other after the load is applied. In other words, the displacements of two adjacent or consecutive points must have identical values (Desai, 1979). Finally, the desired stresses are calculated from the stress–strain relationship. Final stress and strain of the soil is calculated over several iterations, assuming a small fraction of the total stress is present at the soil surface at each iteration. Perumpral *et al.* (1971) assumed an increment of 7 kPa stress at the soil surface during their simulation.

Pollock *et al.* (1984) simulated the multipass effects of vehicles on volumetric strain (soil compaction). Figure 23 shows the differences in volumetric strain in clay and sand after one pass of an 18.4-38 bias-ply tractor tire. As expected, compaction at a given soil depth is greater in clay soils than in sandy soils. These strains can be converted easily to iso-stress lines using the stress–strain relationship. However, field validation tests are needed to check the accuracy of the predicted stress or strain profiles using this model.

### B. Models Based on Stress Distribution

Another approach to predicting soil compaction with depth is to estimate the stress distribution from an applied load and then calculate changes in volumetric strain from the predicted stress. A rather simple way to estimate the stress profile is to assume uniform distribution of stress, $\sigma_z$, in the soil on every horizontal plane under a contact area, $A$, bearing a load, $P$:

$$\sigma_z = P/A \tag{30}$$

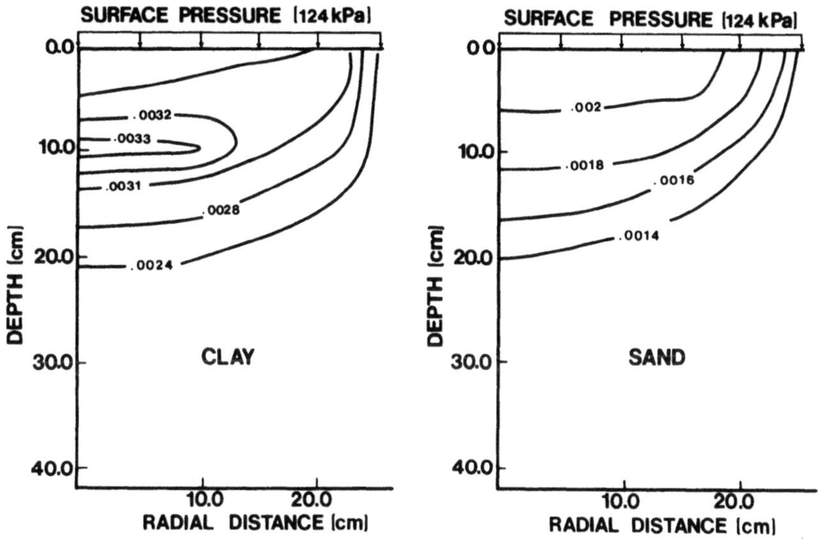


**Figure 23.** Contours of volumetric strain after one pass of an 18.4-38 bias ply tractor tire. From Pollock *et al.* (1984).

This simple assumption of linear stress distribution is not quite true. The stress from a load on the ground surface distributes into the soil while diminishing in intensity, not only vertically with depth, $z$, but also horizontally and away from the axis of loading. Jumikis (1964) demonstrated this concept by considering one row of spheres piled up vertically as in Figure 24. The stress from a load $P = 1$ distributes through the spheres

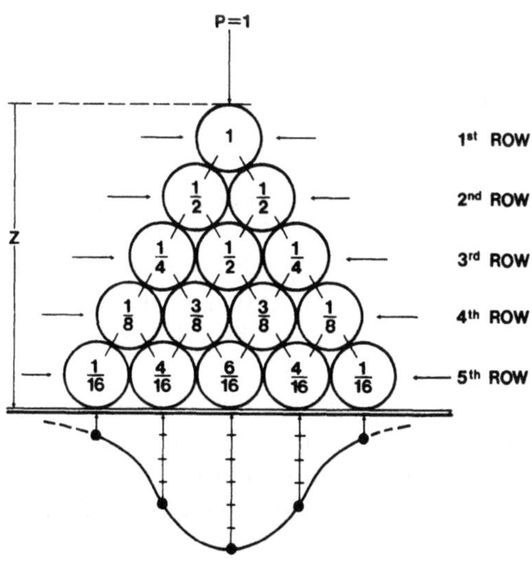


**Figure 24.** Stress distribution through spheres. From Jumikis (1964).

to form an intergranular stress distribution. At a given depth $z$, the stress, viz., reaction to $P = 1$, is larger under the middle sphere in row 5 in the axis of loading than under the sphere next to the middle one. Therefore, compressive stress is larger here than laterally away from the axis of symmetry of loading.

Boussinesq (1885) used the theory of elasticity to predict stress in soil under a point load. He suggested that under the influence of gravity there existed a large enough stress, and also a large enough friction between granular particles, to check any translocation of the granules (Jumikis, 1964). Boussinesq's theory applies to the simplest case of loading a solid, homogeneous, elastic–isotropic, semi-infinite medium. Boussinesq showed that the polar stress $\sigma_R$ in Figure 25 can be expressed as:

$$\sigma_R = (3P/2\pi R^2) \cos \beta \qquad [31]$$

Because the elasticity of soil increases with soil depth, Frohlich (1934) introduced a parameter, $\nu$, defined as the "number of order of stress distribution" or simply, the concentration factor. Thus, equation [31] can be rewritten:

$$\sigma_R = (\nu P/2\pi R^2) \cos^{\nu-2} \beta \qquad [32]$$

Similarly, other stresses at a point in the soil (Figure 25) are as follows (Jumikis, 1964):

1. Vertical normal stress:

$$\sigma_z = (\nu P/2\pi R^2) \cos^{\nu}\beta = (\nu P/2\pi z^2) \cos^{\nu+2} \beta \qquad [33]$$

2. Horizontal, radial normal stress:

$$\sigma_h = (\nu P/2 \pi R^2) \cos^{\nu-2} \beta(\sin^2 \beta) = (\nu P/2 \pi z^2) \cos \beta(\sin^2 \beta) \qquad [34]$$

3. Vertical and horizontal shearing stress:

$$\tau = (\nu P/2 \pi R^2) \cos^{\nu-1} \beta(\sin \beta) = (\nu \mathrm{P}/2 \pi \mathrm{z}^2) \cos^{\nu+1} \beta(\sin \beta) \qquad [35]$$

4. Horizontal, circumferential normal stress (for $m = 2$):

$$\sigma_t = 0 \qquad [36]$$

where $m$ is the Poisson coefficient.

Stresses calculated from equations [31] through [36] correspond to those in an elastic, isotropic body whose Poisson coefficient, $m$, equals 2 if $\nu$ is set equal to 3. Additionally, the above equations are valid in the region where the structure of the soil under the load is undisturbed, i.e., outside the plastic zone.

Figure 26 shows the distribution of the 38 kPa isostress line at various $\nu$ values after a surface application of 181,600 kg. An increase in $\nu$ leads to deeper penetration of a given iso-stress line.

Soehne (1953) used Frohlich's modification of the Boussinesq equation to predict iso-stress lines under a tractor tire. This procedure involves

**Figure 25.** Stresses on a soil element caused by a point load, P. From Soehne (1953).

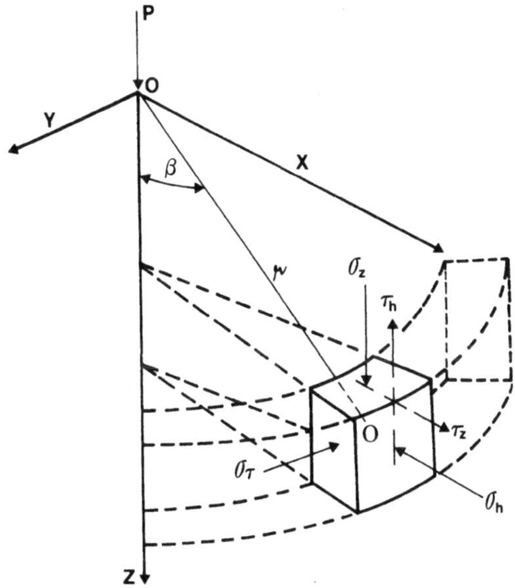


dividing the contact area between tires and ground into 25 load elements (Figure 27) in whose center of gravity a fraction of the surface load is applied as a concentrated load. The main stress ($\sigma_1$) at a point in the soil profile is then calculated from the summation of stresses from the 25 load elements using equations [37] through [41]:

$$\sigma_1 = \sigma_z \cos^2 \alpha + \sigma_x \sin^2 \alpha + \tau_{xz} \sin 2\alpha \qquad [37]$$

$$\sigma_x = \sigma_R \cos^2 \phi \qquad [38]$$

$$\tau_{xz} = \tau \cos^2 \phi \qquad [39]$$

where

$$\cos \phi = (x_i + x)/[(x_i + x)^2 + y_i^2]^{1/2} \qquad [40]$$

and

$$\tan 2\alpha = 2\tau_{xz}/(\sigma_z - \sigma_x)$$

Soehne assumed that soil compaction was mostly caused by $\sigma_1$. Figures 28 and 29 define the variables used in equations [37] and [41]. Soehne (1953) arbitrarily suggested $\nu$ equal to 4, 5, and 6 for hard, firm, and soft conditions of soil, respectively. Soehne's description of firmness applies to a combination of both bulk density and water status of the soil. Blackwell and Soane (1981) showed that Frohlich's equation required different $\nu$ values with soil depth in order to match predicted bulk density with measured values. Gupta and Larson (1982) estimated that the $\nu$ values that gave the best prediction of iso-bulk-density line under laboratory condi-

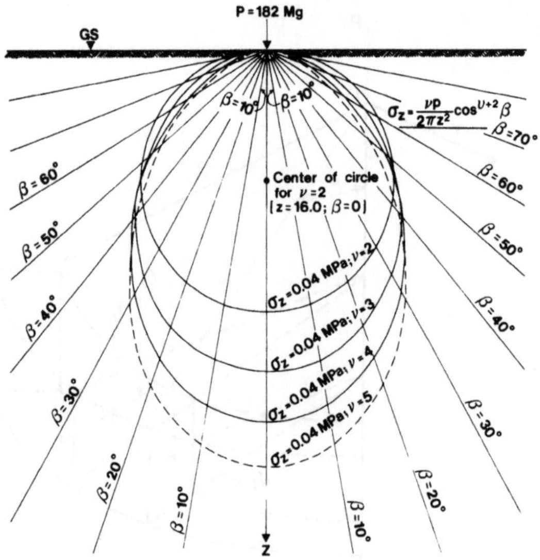


**Figure 26.** Shape of isobars for various Frohlich's stress concentration factors, with constant stress of $\sigma_z = 0.04$ MPa. From Jumikis (1964).

tions were small near the surface and increased with depth. Based on a field experiment, Ram (1984) concluded that (1) $\nu$ varied from 3.28 to 5.84 for sandy and loamy soils in normal field conditions and this factor increased to 7.74 in a low-strength powdered silt loam soil, and (2) $\nu$ increased with an increase in soil moisture content at a constant bulk density and decreased with an increase in bulk density of soil at the same moisture level.

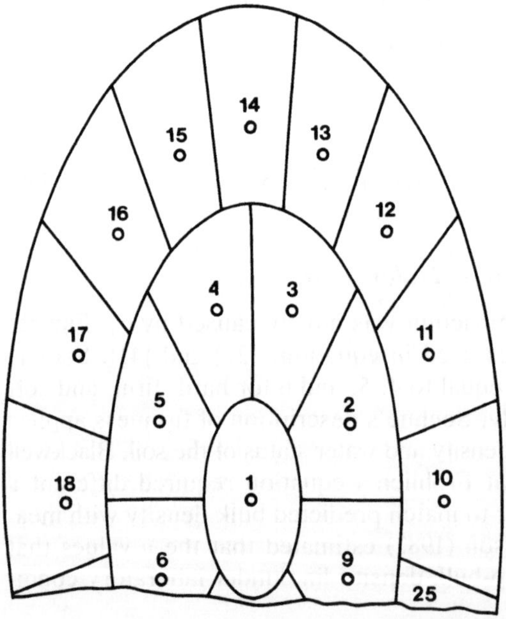


**Figure 27.** A surface load broken up into 25 load elements. Open circles represent the center of gravity with a pressure distribution according to a parabola of 5th potency. From Soehne (1953).

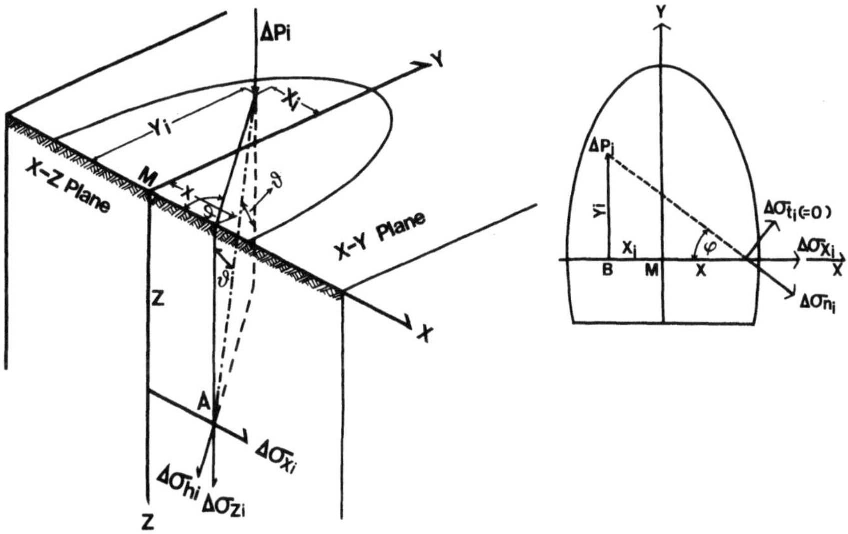


**Figure 28.** Stresses at a point on the *xz* plane as a result of point load at $x_i$, $y_i$ on the soil surface. From Soehne (1953).

Using a combination of two different size tires and two different loads, J.H. Taylor *et al.* (1978) concluded that Frohlich's equations approximately described the soil pressure distribution in both a sandy soil and a clay soil with uniform density profiles. However, when the compacted layer was introduced into the density profile, these equations were no longer suitable. In their study, comparisons of measured and predicted bulk densities or pressures were made at the center of axis under the tire.

Based on the concepts of Soehne (1953), Gupta and Larson (1982) developed a model to predict the distribution of applied stress after passage of an agricultural vehicle. Bulk densities were predicted using the stress–strain relationship obtained in the uniaxial compression test (Larson *et al.*, 1980). Gupta *et al.* (1985b) tested the soil compaction model on Waukegan silt loam (Typic Hapludoll) and Doland loam (Udic Haploboroll) in soil bins at Rosemount and Morris, Minnesota, respectively, and on Nahal Oz silt loam (Calcic Malic Haploxeralf) in Israel. Field tests in Israel covered several moisture contents, two speeds, two groups of passes, and three different tractors. The soil compaction model underpredicted the mechanical stress profile, but the predicted bulk density profiles were close to the measured values. Maximum deviation between predicted and measured bulk densities were 0.12, 0.13, and 0.33 Mg $m^{-3}$ for the Waukegan silt loam, Doland loam, and Nahal Oz loam, respectively. Some of the differences between measured and predicted stress profiles were related to uncertainties of soil–transducer contact and the soil variability in the bins. A sensitivity analysis of the soil compaction model to changes in $\nu$ values did not significantly alter the differences between measured

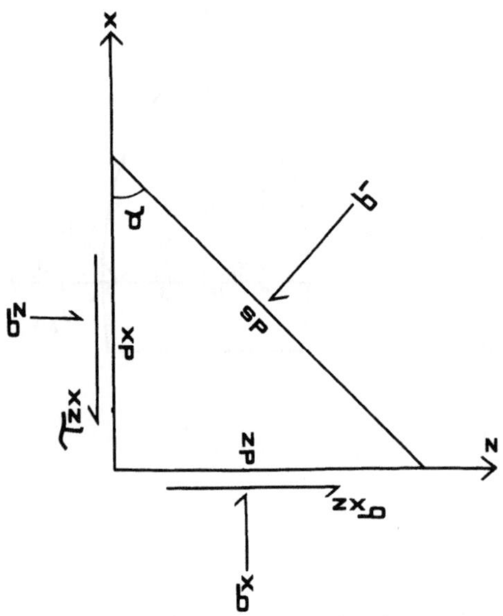


**Figure 29.** Equilibrium of the stresses on a triangular element whose hypotenuse is vertical to the direction of main stress. From Soehne (1953).

and predicted bulk densities on a Waukegan silt loam and Doland loam (Gupta *et al.*, 1985b). Changes in predicted bulk densities from $\nu = 3$ to $\nu = 6$ were within the range of error encountered in the measurement of field densities.

## IX. Application of Field Soil Compaction Models

Field soil compaction models such as that of Gupta and Larson (1982) can be used with soil survey, climatic, and machinery use (load) data bases to (1) delineate geographical areas that are susceptible to excessive compaction, (2) provide guidelines on loads such that excessive compaction does not penetrate below the till layer, (3) delineate areas according to the type of soil–plant limiting conditions, and (4) evaluate the effectiveness of management schemes that prevent or alleviate excessive compaction. The following is an example of a procedure that can be used to achieve the first two applications.

1. A climatic data base can be used to obtain a probability distribution of soil water content by horizons corresponding to the times when heavy machinery is used on the farm. A hypothetical example of this distribution is shown in Figure 30. Fifty percent probability refers to an average soil water content, whereas 75% probability defines wet conditions and 25% probability defines dry conditions.
2. A machinery use data base can be used to define an average load used in a given geographical area. The soil compaction model can then be

**Figure 30.** A hypothetical cumulative probability distribution of soil water content at two soil depths.

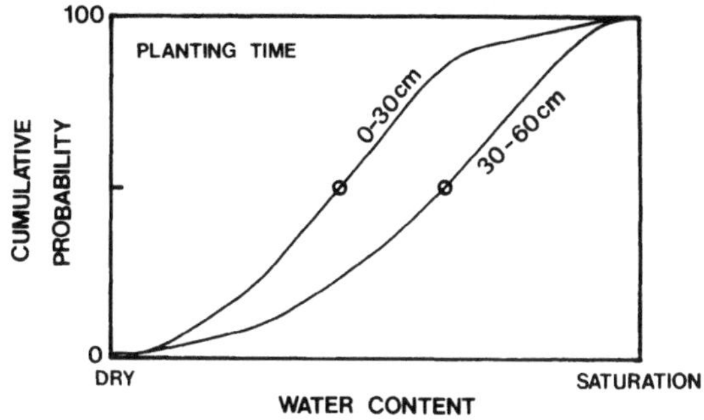


used to predict mechanical stress distribution profiles for an average machinery load and the three ranges of soil water content estimated in step 1. Figure 31, taken from Gupta and Larson (1982), shows an example of such a distribution.

3. Iso-stress lines can be converted into iso-bulk-density lines using the compression data base of Gupta *et al.* (1985a) and the particle-size distribution obtained from the soil survey data base. Using the limits of excessive compaction described earlier, zones of soil and plant limiting conditions can be delineated as shown in Figure 32. This figure shows that (a) at depths of 0 to 9 cm, the applied mechanical stress is higher than the critical value for shearing soil aggregates; (b) at depths of 0 to 18 cm, air-filled porosity is below 10% and therefore critical for gas-

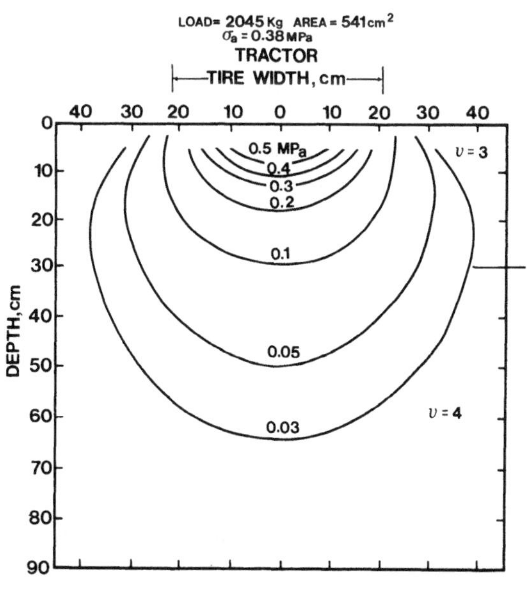


**Figure 31.** Predicted iso-stress lines in a Webster clay loam (Typic Haplaquoll) after a passage by a tractor load.

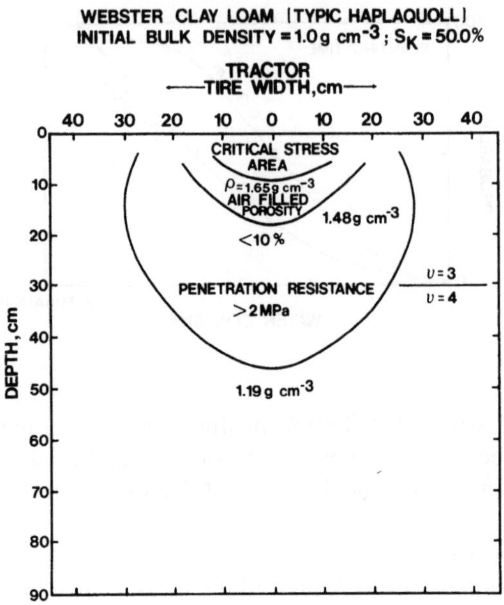


**Figure 32.** Delineation of predicted soil bulk density profile after a passage of a tractor into zones based on (1) stress critical for shearing aggregates, (2) limiting air-filled porosity, and (3) critical penetration resistance.

eous exchange; and (c) at depths of 0 to 45 cm, the penetration resistance is higher than the critical limits (2 MPa) for root growth.

4. The greatest penetration of a limiting condition can be used as a criterion while delineating the susceptibility of soil to excessive compaction for a geographical area. For example, if the limiting conditions do not penetrate below the 30-cm depth (till layer), it could be assigned to category I. Penetration of limiting conditions by each additional 10 cm could be grouped in other categories as shown in Figure 33. Repeating this process for each of the dominating soils in a geographical area, sections can be delineated according to the susceptibility of soil to excessive compaction at various probabilities of soil wetness (Figure 33). Figure 33 shows an expansion of geographical areas where limiting conditions do not penetrate below the till layer with decreasing soil water content.
5. Another type of area delineation is the surface applied stresses, which are not conducive to penetration of most limiting conditions below the till layer. The premise of this assumption is that soil compaction in the till layer is easily manageable. To acheive this objective, we go back to step 2 and simulate iso-stress and iso-bulk-density lines at various applied stresses such that limiting soil conditions do not penetrate below the till layer. A geographical area can then be delineated into subareas according to the categories of surface stresses that are not conducive to the penetration of excessive loads below the till layer. Figure 33

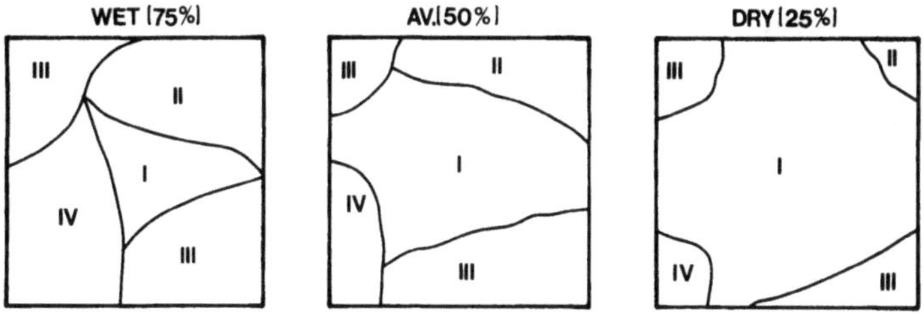


| CATEGORY | DEPTH OF PENETRATION | APPLIED STRESS |
|---|---|---|
| | -cm- | -KPa- |
| I | 20–30 | 1000 |
| II | 30–40 | 750 |
| III | 40–50 | 500 |
| IV | 50–60 | 250 |

**Figure 33.** Delineation of a geographical area according to the susceptibility of soils to excessive compaction at three probability levels of soil wetness below 30 cm depth.

shows an example of these categories. With increasing wetness, the area where higher loads can be applied (without limiting conditions penetrating below the till layer) shrinks.

Another application of this approach is delineating areas according to the limiting conditions. For example, in category I soil the limiting condition that penetrates to the greatest depth is soil resistance, whereas in categories II and III it may be reduced hydraulic conductivity or reduced oxygen diffusion. The scheme described above can also be useful in evaluating management systems that prevent or alleviate excessive compaction.

# X. Areas of Future Research

Developments in soil compaction modeling are far enough along to make use of these models in developing general guidelines for use by extension scientists and farmers. The drawback to the use of these models is the lack of (1) information on variables (including threshold values) that can define excessive compaction, (2) data bases on these limiting variables covering a wide range of soil types and several crops, and (3) information on probabilities of soil water content distribution for a large number of places in a geographical area. Of course, refined calculation and validation of soil compaction models should be a continuous process in soil compaction research.

### Acknowledgment

This work was partially supported by the Binational Agricultural Research Development Fund between the United States and Israel under a grant No. US-337-80.

### References

Akram, M., and W.D. Kemper. 1979. Infiltration of soils as affected by the pressure and water content at the time of compaction. *Soil Sci. Soc. Am. J.* 43:1080–1086.

American Society for Testing and Materials. 1985. Standard test methods for moisture-density relations of soils using 5.5-lb (2.5 kg) hammer and 12-in (304.8 mm) drop. *1985 Annual Book of ASTM Standards*, Part 19. American Society Testing Materials, Philadelphia, PA, pp. 157–162.

Amir, I., G.S.V. Raghavan, E. McKyes, and R.S. Broughton, 1976. Soil compaction as a function of contact pressure and soil moisture content. *Can. Agric. Eng.* 18:54–57.

Bailey, A.C., and G.E. Vanden Berg. 1967. Yielding by compaction and shear in unsaturated soils. *Trans. Am. Soc. Agric. Eng.* 11:307–311, 317.

Bailey, A.C., C.E. Johnson, R.L. Schafer, T.A. Nichols, and R.D. Grisso. 1985. A compaction model for agricultural soil. In: *Soil Dynamics as Related to Tillage Machinery Systems. Proc. International Conference on Soil Dynamics.* Vol. 2. June 17–19, 1985 National Tillage Machinery Laboratory and the Agricultural Engineering Department, Alabama Experiment Station, Auburn University, Auburn, AL. pp. 203–212.

Baver, L.D., W.H. Gardner, and W.R. Gardner. 1972. *Soil Physics*. John Wiley and Sons, New York.

Bishop, A.W. and D.J. Henkel. 1962. The measurement of soil properties in the triaxial test. Edward Arnold, London, 228 pp.

Blackwell, P.D., and B.D. Soane. 1981. A method of predicting bulk density changes in field soils resulting from compaction by agricultural traffic. *J. Soil Sci.* 32:51–65.

Boussinesq, J. 1885. *Application des Potentiels a l'Etude de l'Equilibre et du Mouvement des Solides Elastiques*. Gauthier-Villais, Paris.

Bowen, H.D. 1981. Alleviating mechanical impedance. In: G.F. Arkin and H.M. Taylor (eds.) *Modifying the Root Environment To Reduce Crop Stress*. American Society of Agricultural Engineering, Monograph No. 4, pp. 21–59.

Bradford, J.M., and S.C. Gupta. 1986. Soil compressibility. In: A. Klute (ed.) *Methods of Soil Analysis,* Part 1: *Agronomy*. (2nd ed.) American *Society of Agronomy, Madison, WI, pp. 479–492.*

*Coleman, G.E., and J.V. Perumpral. 1974. The finite element analysis of soil compaction. Trans. Am. Soc. Agric. Eng.* 17:856–860.

Desai, C.S. 1979. *Elementary Finite Element Method*. Prentice-Hall, Englewood Cliffs, NJ.

Dexter, A.R., and D.W. Tanner. 1973. The response of unsaturated soils to isotropic stress. *J. Soil Sci.* 24:491–502.

Frolich, O.K. 1934. Druckverteilung im Baugrunde. Verlag von Julius Springer, Wien.

Gerard, C.J., P. Sexton, and G. Shaw. 1982. Physical factors influencing soil strength and root growth. *Agron. J.* 74:875–879.

Gill, W.R. 1971. Economic assessment of soil compaction. In: K.K. Barnes, W.M. Carleton, H.M. Taylor, R.I. Throckmorton, and G.E. Vanden Berg (eds.) *Compaction of Agricultural Soils*. American Society of Agricultural Engineering, Monograph No. 1, St. Joseph, MI, pp. 431–458.

Grable, A.R., and E.G. Siemer. 1968. Effects of bulk density, aggregate size, and soil water suction on oxygen diffusion, redox potentials, and elongation of corn roots. *Soil Sci. Soc. Am. Proc.* 32:180–186.

Gupta, S.C., and W.E. Larson. 1982. Modeling soil mechanical behavior during tillage. *In:* P. Unger, D.M. Van Doren Jr, F.D. Whisler, and E.L. Skidmore (eds.), *Symposium on Predicting Tillage Effects on Soil Physical Properties and Processes*. American Society of Agronomy, Spec., Publ. 44, Madison, WI, pp. 151–178.

Gupta, S.C., A. Hadas, W.B. Voorhees, D. Wolf, W.E. Larson, and E.C. Schneider. 1985a. Development of guides for estimating the ease of compaction of world soils. Res. Rep. submitted to BARD Fund on completion of Project Grant No. US-337-80.

Gupta, S.C., A. Hadas, W.B. Voorhees, D. Wolf, W.E. Larson, and E.C. Schneider. 1985b. Field testing of a soil compaction model. In: *Traction and Transport as Related to Cropping Systems. Proc. Int. Conf. Soil Dynam.*, Vol. 5. National Tillage Machinery Laboratory and the Agricultural Engineering Department, Alabama Experiment Station, Auburn University, Auburn, AL. pp. 979–994.

Hillel, D. 1971. *Soil and Water*. Academic Press, New York.

Jumikis, A.R. 1964. *Mechanics of Soils*. Van Nostrand, Princeton, NJ.

Koolen, A.J., and H. Kuipers. 1983. *Agricultural Soil Mechanics*. Springer-Verlag, New York.

Larson, W.E., and S.C. Gupta. 1980. Estimating critical stresses in unsaturated soils from changes in pore water pressure during confined compression. *Soil Sci. Soc. Am. J.* 44:1127–1132.

Larson, W.E., S.C. Gupta, R.A. Useche. 1980. Compression of agricultural soils from eight soil orders. *Soil Sci. Soc. Am. J.* 44:450–457.

McKibben, E.G. 1971. Introduction. In: K.K. Barnes, W.M. Carleton, H.M. Taylor, R.I. Throckmorton, and G.E. Vanden Berg (eds.) *Compaction of Agricultural Soils*. American Society of Agricultural Engineering, Monograph No. 1, St. Joseph, MI. pp. 3–6.

Perumpral, J.V., J.B. Liljedahl, and W.H. Perloff. 1971. The finite element method for predicting stress distribution and soil deformation under a tractive device. *Trans. Am. Soc. Agric. Eng.* 14:1184–1188.

Pollock, D., Jr., J.V. Perumpral, and T. Kuppusamy. 1984. Multipass effects of vehicle on soil compaction. Paper No. 84–1054 presented at the winter meetings of the American Society of Agricultural Engineers, Chicago, IL.

Ram, R.B. 1984. Pressure measurement in the soil under the load. *Soil Till. Res.* 4:137–145.

Russell, R.S., and M.J. Goss. 1974. Physical aspects of soil fertility: The response of roots to mechanical impedance. *Neth. J. Agric. Sci.* 22:305–318.

Soehne, W.H. 1953. Pressure distribution in the soil and soil deformation under tractor tires. *Grundl. Landtech.* 5:49–63.

Soehne, W.H. 1976. Terramechanics and its influence on the concept of tractors, tractor power development and energy consumption. *J. Terramech.* 13:27–43.

Stone, J.A., and W.E. Larson. 1980. Rebound of five one-dimensionally compressed unsaturated granular soils. *Soil Sci. Soc. Am. J.* 44:819–822.

Taylor, H.M., G.M. Robertson, and J.J. Parker, Jr. 1966. Soil strength-root penetration relations for medium to coarse-textured soil materials. *Soil Sci.* 102:18–22.

Taylor, J.H., E.C. Burt, and A.C. Bailey. 1978. Traction and compacton of big tractors. Paper No. 78-1029, presented at the American Society of Agricultural Engineers, St. Joseph, MI.

Unger, P.W., H.V. Eck, and J.T. Musick. 1981. Alleviating plant water stress. In: G.F. Arkin and H.M. Taylor (eds.) *Modifying the Root Environment To Reduce Crop Stress*. American Society of Agricultural Engineers Monograph No. 4, pp. 61–97.

Voorhees, W.B., D.A. Farrell, and W.E. Larson. 1975. Soil strength and aeration effects on root elongation. *Soil Sci. Soc. Am. J.* 39:948–953.

Voorhees, W.B., C.G. Senst, and W.W. Nelson. 1978. Compaction and soil structure modifications by wheel traffic in the northern Corn Belt. *Soil Sci. Soc. Am. J.* 42:344–349.

Wolf, D., and A. Hadas. 1984. Soil compaction effects on cotton emergence. *Trans. Am. Soc. Agric. Eng.* 27:655–659.

# Sources, Amounts, and Forms of Alkali Elements in Soils*

A.D. Scott and S.J. Smith

I. Introduction ........ 101
II. Differentiating Characteristics ........ 102
A. Significance in Soils ........ 103
B. Chemical Properties ........ 105
III. Geochemistry ........ 107
IV. Mineralogy ........ 113
V. Soil Content ........ 117
A. Range of Content and Geographical Distribution ........ 117
B. Particle Size ........ 120
C. Profile Distribution ........ 121
D. Fertilizer, Organic Matter, and Precipitation Effects ........ 122
E. Artificial Isotopes ........ 123
VI. Soil Forms ........ 124
A. Soluble ........ 125
B. Exchangeable ........ 130
C. Nonexchangeable ........ 135
D. Fixed ........ 136
References ........ 138

## I. Introduction

The elements Li, Na, K, Rb, Cs, and Fr constitute group 1A of the periodic table and are collectively referred to as the alkali metals or alkalies. Stable isotopes of all these elements except Fr exist in soils; they are $^{6}Li$, $^{7}Li$, $^{23}Na$, $^{39}K$, $^{41}K$, $^{85}Rb$, and $^{133}Cs$. Of the various radioactive isotopes of the alkali metals, the artificial isotope $^{137}Cs$ and the natural isotopes $^{40}K$ and $^{87}Rb$ are the main ones found in soils. There are several isotopes of Fr but only the radioactive isotope $^{223}Fr$ from the $\alpha$ decay of Ac occurs nat-

*Journal Paper No. J-12134 of the Iowa Agriculture Home Economics Experiment Station, Ames, Iowa. Project No. 2692.

urally. Inasmuch as only 1.2% of the Ac undergoing radioactive decay yields Fr and the half-life of this isotope of Fr is only 21 min, however, even the existence of Fr in soils has been difficult to establish. Some work on the geochemistry of Fr and the contribution of Fr to the radioactivity of soils has been reported (Maddock, 1963), but this element is not considered further here.

The literature pertaining to the alkali elements is voluminous but includes many texts and reviews that provide a general background for this discussion. The chemistry of the alkali elements has been considered in great detail by Mellor (1922) and by others in supplements to Mellor's *Comprehensive Treatise on Inorganic and Theoretical Chemistry* (Briscoe *et al.*, 1961, 1963). Sections of these supplements deal with the occurrence and biological properties of each alkali element and consider the alkalies in relation to soils. A more recent book by Bowen (1979) emphasizes an environmental approach to the chemistry of the alkalies and other elements. For a discussion of the geochemistry of the alkali elements, reference can be made to several texts, but those of Rankama and Sahama (1950), Goldschmidt (1954), and Mason (1966) are particularly useful. There are also reviews that pertain specifically to the geochemistry (Gentili, 1954; Heier and Adams, 1964; Vlasov, 1966a; Heier and Billings, 1970) and mineralogy (Vlasov, 1966b; International Potash Institute, 1972) of the alkali elements. For those aspects about the alkalies that involve their existence in soils, attention must be given to a wide variety of publications that usually emphasize individual elements and specific topics. Nevertheless, comprehensive reviews exist for Li (Bradford, 1966; Wells and Whitten, 1972; Angino *et al.*, 1974); Rb (Diest and Talibudeen, 1967; Coughtrey and Thorne, 1983a), and Cs (Davis, 1963; Coughtrey and Thorne, 1983b), individually or together (Vinogradov, 1959; Aubert and Pinta, 1977; Kabata-Pendias and Pendias, 1984); Na (Kelley, 1951; US Salinity Laboratory Staff, 1954; Tinker, 1967a; Bresler *et al.*, 1982; Coughtrey *et al.*, 1983); and K (Reitemeier, 1951; Richter, 1965; Duthion, 1968; Boyer, 1972; Su, 1978; K.D. Ritchie, 1979; Metson, 1980).

## II. Differentiating Characteristics

Marked differences exist among the alkali elements in regard to their relative abundance in soils (e.g., atom ratio K/Cs $\sim 10^4$ in average soil). The significance of such differences, however, is tempered by equally large differences among the alkalies in regard to why they are important in soils and the concentration levels at which their roles are fulfilled. Also, the sources of the alkalies in soils vary with a few chemical characteristics of the elements that govern their inclusion in mineral crystals and their adsorption by clay materials. Thus, to provide a perspective for the specifics on sources, amounts, and forms of the alkalies in soils, a short account of the characteristics that determine their importance and distribution in soil is given first.

### A. Significance in Soils

Because Li is not an essential plant nutrient and soils generally contain much less than 100 mg $kg^{-1}$ total Li, little interest in soil Li may be anticipated. According to Bradford (1966), Call (1961), and Kabata-Pendias and Pendias (1984), however, this is not true because very little Li can produce stimulating or toxic effects in plants. Of the alkali elements, Li is the most toxic to plants and is sufficiently so that even indigenous levels of Li in soil have proved to be toxic to citrus. Lithium excess for crops can also arise from irrigation water and soil management practices. In sand-culture experiments little more than 5 mg Li $L^{-1}$ in solution can be enough to be toxic to some plants. On the other hand, beneficial effects of small amounts of Li on plant growth have been reported for several years, and the possibility of potential effects of Li on human health has focused attention on Li in ground water (Angino *et al.*, 1974). Considering the therapeutic value of Li (Johnson, 1984), even more emphasis on soil sources of Li in the food chain can be expected.

Only a few plants seem to require Na, but there are many crops that are benefited by this element (Bear, 1953; Coughtrey *et al.*, 1983). Some crops (e.g., sugar beet) respond to large applications of Na salts (Tinker, 1967b); others develop toxicity symptoms with a relatively low ($<10\%$) exchangeable sodium percentage (Pearson, 1960). Such a combination of observations has led to a prolonged controversy over the value of Na as a nutrient to plants (Tinker, 1967a) and the continuing efforts to verify the essentiality and role of Na in plants (Brownell, 1979). On the other hand, it is well established that undesirable physical properties of the soil can be produced by exchangeable Na and accumulations of Na salts can be a major contributor to the development of salinity in soils of arid and semiarid regions (Bresler *et al.*, 1982). Therefore, much of the concern about Na centers on the deleterious effects of too much Na in soils as well as its role in plant nutrition.

Being the only alkali element that is a major plant nutrient, K has been given far more attention than any other alkali element in soils. Potassium is generally abundant in soils, but the rate and degree of K release to plant-available forms and the need and use of fertilizer additions for particular cropping conditions are ever-present problems. Also, through its association with feldspars and micaceous minerals that abound in soils, K is involved in various processes of weathering and clay mineral behavior. The importance of these aspects of soil K is evident in the numerous publications on the role of soil K in plant growth, the chemical and mineralogical behavior of K in soils, and the production and use of K fertilizers (Rich, 1968; Sawhney, 1972; Grimme and Nemeth, 1978; Schroeder, 1978; Mengel and Kirby, 1980; Barber, 1984; H.W. Martin and Sparks, 1985; Tisdale *et al.*, 1985). In addition, the naturally occurring radioactive isotope $^{40}K$ has been used extensively for dating geological materials (Heier and Adams, 1964). According to Talibudeen (1964), two thirds of the radioactivity in surface soils comes from K. Agronomic research has also made use of this $^{40}K$ as well as the artificial isotope $^{42}K$ (Bouat, 1969). Additional details about the use of the various isotopes of K, Rb, and Cs as tracers in soil fertility and erosion studies have been described in the recent review by Menzel and Smith (1984).

Interest in Rb in soils stems mainly from the potential use of Rb or the radioactive isotope $^{86}Rb$ as a tracer for K in studies involving soil processes and plant uptake (Diest and Talibudeen, 1967; Eckert and McLean, 1980; Coughtrey and Thorne, 1983a). Therefore, the forms and reactions of Rb in soils and their relation to K are of prime concern. Because of the similarity between K and Rb in regard to cation size and chemical character, there is a close association or coherence of the elements in minerals and rocks, which has led to frequent use of the ratios of indigenous amounts of Rb and K in geochemical studies (Taylor, 1965; Heier and Billings, 1970). Additional interest in the indigenous Rb in soils stems from the use of $^{87}Rb$ in dating geological materials (G.R. Martin, 1963). In surface soils, about 7% of the natural radioactivity is from this $^{87}Rb$ (Talibudeen, 1964). Soil contamination with $^{86}Rb$ from radioactive fallout has received some attention (Frere *et al.*, 1963), but the short half-life of this isotope (18.6 days) reduces the importance of this aspect of Rb in soils.

The scarcity of indigenous Cs in soils and minerals has created analytical difficulties that have limited the work done on this naturally occurring portion of the element in soils. Cesium is more toxic to plants than Rb but less than Li and is enough different from K in size and chemical behavior to make the use of K/Cs ratios in minerals or the use of Cs as a tracer for K very limited. Indeed, Souty *et al.* (1975) determined the uptake of K and $^{137}Cs$ by plants grown in nutrient solution and concluded the Cs could not be used as a quantitative tracer of K. Nevertheless, interest in the reactions of Cs with soils and the uptake of Cs by plants exists because radioactive isotopes of Cs from nuclear tests have contributed to the hazards of fallout in soils (Davis, 1963; Frere *et al.*, 1963). In general, Cs is not a very mobile element in soils, but Evans *et al.* (1983) have shown that some soil conditions can enhance the mobility and biological availability of Cs and thereby the danger of human assimilation of $^{137}Cs$. The same property of Cs retention by soils has led to the recent use of fallout $^{137}Cs$ as a tracer of the source and extent of soil movement and loss (Ritchie *et al.*, 1970; McCallan *et al.*, 1980; McHenry and Bubenzer, 1985). The fact that $^{137}Cs$ is an abundant component of radioactive wastes has also focused attention on Cs–soil relationships because aluminosilicates play a role in the waste management (Robinson, 1962; Lee and Tank, 1985).

## B. Chemical Properties

The alkali elements are very much alike in that they all have the same outer electron configuration, form monovalent cations readily, are very reactive, form very soluble compounds, etc. They are also alike in being classified geochemically as oxyphile or lithophile elements because they are generally bonded to oxygen and concentrated in the Earth's crust as silicates. Still, the alkali elements are distinctive in regard to their exchange and crystallochemical behavior. As a result, element separations occurred during the formation and alteration of rocks and soils and specific distributions of each element in minerals and soils have evolved. Moreover, these differences in alkali behavior are evident in soils today. Explanations for these differences generally involve the characteristics of the elements listed in Table 1.

Relative to many other cations, the univalent character of the alkali cations is a major distinguishing feature because cations with higher charge but comparable size are preferred in mineral structures that accept either cation. Compared to one another, however, the alkali elements differ mainly with respect to ionic size and their tendency to lose electrons and form ionic bonds. Even though ionic size is an operational concept that provides definite values for specific conditions (Stern and Adams, 1959), there are several tables of ionic radii in common use (see compilations by Heydemann, 1969). Considering the crude approximation made in

**Table 1.** Characteristics of alkali elements

| | Li | Na | K | Rb | Cs |
|---|---|---|---|---|---|
| Atomic number[a] | 3 | 11 | 19 | 37 | 55 |
| Atomic weight[a] | 6.939 | 22.9898 | 39.102 | 85.49 | 132.905 |
| Isotopes [abundance, %][a] | 6 [7.4]<br>7 [92.6] | 23 [100] | 39 [93.1]<br>40 [0.01]<br>41 [6.9] | 85 [72.2]<br>87 [27.8] | 133 [100] |
| Crystal radii (nm)[a] | 0.076[b] | 0.098 | 0.133 | 0.149 | 0.165 |
| Ionic potential[c] | 1.32 | 1.02 | 0.75 | 0.68 | 0.60 |
| Ionization potential (ev)[a] | 5.363 | 5.12 | 4.318 | 4.159 | 3.87 |
| Electronegativity[d] | 0.95 | 0.9 | 0.8 | 0.8 | 0.75 |
| Hydrated radii (nm)[e] | 0.340 | 0.276 | 0.232 | 0.228 | 0.228 |
| Hydration energy (kJ $mol^{-1}$)[f] | 519 | 406 | 322 | 293 | 264 |
| Polarizability × $10^{24}$ ($cm^3$)[f] | 0.08 | 0.2 | 0.83 | 1.39 | 2.42 |

[a]Heydemann (1969).
[b]Donnay and Gryder (1964).
[c]Valence/crystal radius.
[d]Gordy and Thomas (1956).
[e]Cotton and Wilkinson (1980).
[f]Shainberg and Kemper (1966).

treating ions as rigid spheres with specific radii, any of the values in these tables, except possibly those for Li, should be adequate to delineate the relative size of the alkali cations. Goldschmidt (1954) originally assigned Li a radius of 0.078 nm but later changed it to 0.070 nm, whereas others have used values of 0.060, 0.068, and 0.070 nm. For Table 1, the Donnay and Gryder (1964) value of 0.076 nm for Li and those of Zemann (from Heydemann, 1969) for the rest of the alkali cations in sixfold coordination have been used. The compatability between these ionic radii and the space available in specific mineral structures is one of the factors governing the distribution of the alkalies among the minerals.

The significance of the chemical bond as a factor controlling the geochemical distribution of elements and the use of ionic potentials (valence/radius), ionization potentials, and electronegativity values as a guide to the nature of the bonds have been reviewed by Ahrens (1964). In the case of the minerals containing alkali elements, the alkalies are bonded mainly to oxygen and by bonds that are primarily ionic. However, the decrease in ionic potential and ionization potential shows there is a decrease in the polarizing ability of the cation and an increase in the ionic character of the bonds formed by the alkalies as the size of the cation increases in the sequence Li < Na < K < Rb < Cs. The electronegativity values yield the same but independent conclusion that the degree of ionic bonding between the alkalies and oxygen increases in the order Li < Na < K < Rb < Cs because oxygen has an electronegativity value of 3.5 and the larger the electronegativity difference between the bonding elements, the greater is the degree of ionic character of the bond.

During periods of sedimentation and soil development the alkali cations go into solution and are redistributed in accordance with their behavior as exchangeable ions. Once they are in solution, the cations tend to hydrate as well as be adsorbed on cation-exchange sites accessible to them. Therefore, various means of assessing the hydration tendency of the cations and the effects of this hydration tendency on the exchangeability of the cations have been used. The hydrated radii given in Table 1 indicate the effective size the cations may achieve in water. They can be used to calculate field functions (ionization potentials/hydrated radius) that give a measure of the electrostatic potential at the surface of the hydrated ion and thus the relative strength with which the hydrated ion may be held on exchange sites (Heier and Adams, 1964). More often, ionic potentials calculated with crystal radii (Mason, 1966) or with hydrated radii (Welby, 1958) are used to account for the effects of ion hydration. Each of these approaches provides a qualitative prediction that the alkali cations will be held with increasing strength in the order Li < Na < K < Rb < Cs. That is, cations with larger crystal radii tend to be less hydrated, have smaller hydrated radii, and be selectively adsorbed because of the closer approach of charges. Several instances where this order for the selective retention of alkali cations on exchange materials has been observed are given by Wiklander (1964).

The effects of cation hydration on the strength with which cations are held on exchange sites can be expressed in quantitative terms by using hydration energy values. By doing so and including the anionic field strength of the exchange sites in his calculations, Eisenman (1962, 1969) has shown that many selectivity sequences are possible for the alkali cations. The normal sequence of cation adsorption preference (Cs > Rb > K > Na > Li) anticipated from differences in their hydration energies seems to apply to montmorillonite, but other sequences obviously can occur with other exchangers.

Many (Goldschmidt, 1954; Welby, 1958; Helfferich, 1962; Heier and Adams, 1964) contend that differences in the polarizability of the cations contribute to their selective retention. Cation polarization increases the strength with which cations are held on exchange sites. Therefore, the increase in cation polarizability shown in Table 1 for the sequence Li < Na < K < Rb < Cs would augment the differences in alkali cation selectivity already predicted by a consideration of coulombic forces and hydration energies. By including the effects of cation polarizability, Shainberg and Kemper (1966, 1967) found the quantitative aspects of Li, Na, and K exchange on montmorillonite could be accounted for. The selectivity with which Cs is retained by montmorillonite, however, exceeds that predicted by such calculations.

For a detailed discussion of the selectivity with which alkali cations are retained on various exchange materials, reference should be made to reviews by Diamond and Wright (1969), Reichenberg (1966), and Goulding (1983). The hydration of alkali cations is considered by Rosseinsky (1965).

# III. Geochemistry

The sources and amounts of the alkali elements in soils are ultimately determined by the processes of soil development but are conditioned by the composition of the solar system and the differentiation of elements within the earth. Therefore, much of the information about the occurrence of alkalies in soils must come from geochemistry.

The origin of the alkali elements is still a matter of conjecture (Burbidge *et al.*, 1957; Day, 1964), but, whatever the process, marked differences in their abundances in the solar system are now known to exist. To obtain this information, various compilations of terrestrial, solar, stellar, and particularly meteoritic data on element abundances have been made and selected data from these sources have been combined in various ways to obtain cosmic, solar system, and whole-Earth abundance values (Clarke, 1924; Aller, 1961; Heier and Adams, 1964; Urey, 1967; Goles, 1969). Various estimates of the abundance of elements in the Earth's crust and in soils have also been made (Vinogradov, 1959; Mason, 1966; R.L. Parker, 1967). A comparison of such data is presented in Table 2 using the convention of normalizing the abundance data in terms of $10^6$ Si atoms (Suess and Urey, 1956) to show the effects of primordial composition and subsequent processes on the relative abundance of the alkalies in soils. To do so with soils, an average Si content of $3.5 \times 10^5$ mg $kg^{-1}$ (average of soils with 60 to 90% $SiO_2$; Kilmer, 1965) was employed.

In the solar system, Na is by far the most abundant alkali element, the order of decreasing abundance is Na > K > Li > Rb > Cs, and the range in abundance is very large (Na/Cs being $>10^5$). Obviously, Li does not follow the general trend of lighter elements being more abundant. To account for the scarcity of Li, Rankama and Sahama (1950) have stressed the possibility of Li conversion to other elements in stellar interiors. Burbidge *et al.* (1957) related the low abundance of Li to a different type of synthesis process. In any event, it is evident from Table 2 that the relative scarcity of Li, Rb, and Cs in soils stems from a lack of these elements in the primordial materials of the Earth.

The differentiation of elements that occurred during the development of the Earth's crust and soil caused some change in the abundance of each alkali element relative to Si and to one another. As a result, the order of decreasing alkali abundance (numbers of atoms) in soils is K > Na > Li > Rb > Cs. Soils are also enriched in Li, Rb, and Cs by comparison with the solar system as a whole, but the numbers of Li, Rb, and Cs atoms in soils are still relatively small.

Details regarding the fate of the alkali elements during the various stages of igneous rock formation and sedimentation (defined as a geochemical process that includes changes in the sediments as well as their origin, transport, etc.; American Geological Institute, 1976) are given in several texts and reviews of geochemistry. Various compilations of geochemical data pertaining to the alkalies are also available. Therefore, only the general

**Table 2.** Relative abundance of alkali elements (atoms per $10^6$ Si atoms)

| | Solar system[a] | Earth's crust[b] | Soil[c] |
|---|---|---|---|
| Li | 16 | 290 | 346 |
| Na | 35,000 | 126,000 | 21,900 |
| K | 2,100 | 67,000 | 28,600 |
| Rb | 4.1 | 107 | 141 |
| Cs | 0.21 | 2.3 | 3 |

[a]Goles (1969).
[b]Calculated, Mason (1966).
[c]Calculated, Vinogradov (1959).

nature of the main processes governing the distribution of the alkalies within the Earth's crust and average levels of each element in crustal rocks and soils are considered here.

When the rock melt (magma) cooled to form the igneous rocks that constitute most of the Earth's crust, a selective crystallization and gravitational separation of minerals occurred; that is, a sequence of silicate minerals formed in the cooling magma and settled in accordance with their melting points and densities, respectively. Because of their tendency to combine with oxygen, the alkalies separated with the silicates, concentrated in the upper layers of the Earth. The distribution of the alkalies in the Earth's crust was established by the degree to which each alkali was included in the silicate minerals that crystallized during the various magmatic stages. Because the formation of the silicates followed a sequence:

anorthite — albite —\
K-feldspar — muscovite,\
olivine — pyroxene — amphibole — biotite —

the space and unbalanced charge available for alkalies in the mineral structures varied with the stage of crystallization. The possibility of a particular alkali cation occupying these available spaces depended upon its valence, size, degree of ionic bonding with oxygen, and concentration in the magma. As the stage of crystallization advanced, cations remaining in the magma became more competitive for sites because their concentrations increased and the mobility of the ions decreased. Because the later crystallates were also less dense, the alkalies separating during the later stages exhibit the greatest degree of enrichment toward the surface of the Earth.

During sedimentation, the processes of weathering, erosion, deposition, and diagenesis are all operative (Mason, 1966). Therefore, the redistribution of the alkalies imposed by sedimentation has a less predictable pattern than that found in igneous rocks, where chemical equilibrium processes were more involved (Spencer *et al.*, 1968). By comparison with the conditions prevailing during the formation of igneous rocks, the sedimentary processes are carried out at low temperature and pressure. Thus, one of the factors governing the redistribution of the alkalies is the stability of the igneous rocks to these changes in environment. Some transport of the alkalies occurs in particulate matter (Sreekumaran *et al.*, 1968) but the dissolution of the alkalies, the exchange behavior of the cations in solution, and the incorporation of these elements in the structures of newly formed minerals are mainly responsible for the differentiation of alkalies in sediments. The main result is a distribution of alkali elements that is closely associated with the clay in the sediments.

The differentiation of alkalies arising from the formation of igneous and sedimentary rocks has been established by numerous rock analyses. The information on Li, Rb, and particularly Cs is still uncertain because the scarcity of these elements has created analytical problems and limited the number of determinations. However, it is evident from the reviews by Angino *et al.* (1974) and Wells and Whitten (1972) that this situation is improving with Li. Moreover, the problem has been partially circumvented for Rb and to a lesser degree for Cs by using reliable K data in conjunction with established K/Rb and K/Cs ratios. Additional difficulties arise in the use of average values for the alkali content of rocks and minerals because the range of content is often very large. The problem is well illustrated by the histograms of Heier and Adams (1964) and their specific example of 116 biotites with Rb contents varying from 0 to 2400 mg $kg^{-1}$. As pointed out by Richards (1956), even sea water is not uniform in alkali content content and is hard to characterize by averages. Nevertheless, average values have been compiled in Table 3 to indicate the general level of the alkalies in different parts of the biosphere, which is defined by Kabata-Pendias and Pendias (1984) as consisting of the surficial part of the lithosphere, the lower part of the atmosphere, and the hydrosphere.

It is obvious from the discussion in Heier and Billings (1970) that there is much uncertainty about the whole-Earth values in Table 3, but they do indicate a probable composition for the magma and thereby provide a basis for identifying the alkali enrichment of various parts of the Earth.

**Table 3.** Average concentration of alkalies in portions of the biosphere[a]

| | Li | Na | K | Rb | Cs |
|---|---|---|---|---|---|
| Whole Earth (mg $kg^{-1}$) | 2 | 4,800 | 640 | 1.6 | 0.06 |
| Continental Earth crust (mg $kg^{-1}$) | 26 | 28,000 | 26,000 | 101 | 3.2 |
| Igneous rocks | | | | | |
| Acidic (mg $kg^{-1}$) | 43 | 27,000 | 33,000 | 165 | 6 |
| Basic (mg $kg^{-1}$) | 13 | 19,000 | 7,500 | 33 | 1 |
| Sedimentary rocks | | | | | |
| Shale (mg $kg^{-1}$) | 62 | 9,600 | 26,000 | 134 | 8 |
| Sandstone (mg $kg^{-1}$) | 15 | 7,000 | 14,000 | 82 | 0.5 |
| Limestone (mg $kg^{-1}$) | 12 | 2,000 | 3,000 | 3 | 0.5 |
| Soils (mg $kg^{-1}$) | 26 | 6,200 | 14,000 | 140 | 5 |
| Stream water (mg $L^{-1}$) | 3[b] | 6.3 | 2.3 | 1[b] | 0.02 |
| Sea water (mg $L^{-1}$) | 170[b] | 10,800 | 392 | 120[b] | 0.3[b] |
| Watershed runoff (mg $L^{-1}$) | — | 1.2 | 6.8 | — | — |
| Ground water (mg $L^{-1}$)[c] | — | 114 | 4.8 | — | — |
| Rainfall (mg $L^{-1}$) | — | 1.4 | 0.3 | — | — |

[a]Based on data of Heier and Adams (1964), Heier and Billings (1970), Mason (1966), Sharpley *et a* (1985), Timmons and Holt (1977), Turekian (1969), Vlasov (1966a), Wedepohl (1969a, b).

[b]μg $L^{-1}$.

[c]Estimated from various US Geological Survey Reports.

Even Na, which is the most abundant alkali in the Earth in terms of weight or atom numbers, accounts for only a few tenths of a percent of the Earth.

The various methods that have been used to determine the composition of the Earth's crust and many of the results of these determinations have been reviewed by R.L. Parker (1967). Included in these results are several that were selected by Mason (1966) and are now given in Table 3. According to Heier and Billings (1970) the values for Li, K, and Rb may be more like 13, 16,000 and 48 mg $kg^{-1}$, respectively. Still, a comparison of the whole-Earth and crust data shows that each of the alkalies is concentrated in the crust. The degree of the crustal enrichment varied from a 10-fold increase in Na to a 700-fold increase in Cs.

Because igneous rocks constitute about 95% of the Earth's crust, the average composition of igneous rocks is usually considered to be the same as the crust. However, the igneous rocks that formed at different stages of magmatic crystallization differ in mineralogy and composition. To show the differences that exist in their alkali content, the rock groups designated as basic (gabbro, basalt) and acidic (granite, rhyolite) on the basis of their silica content have been compared in Table 3. The concentration of alkali elements in the early formed (basic) igneous rocks is quite low. Competition from Mg and Fe cations for the same structural positions in silicate minerals is responsible for the lack of Li. As the concentration of Li in the magma increased and the magma cooled, progressively more Li was incorporated in succeeding crystallates and the Li/Mg ratio for the rocks increased. Sodium had to compete with Ca for space in the plagioclase crystals, but even the plagioclase that crystallized at the beginning of the main magmatic stage contained some Na because the concentration of Na in the magma was very high. In addition, the early forming pyroxenes and amphiboles incorporated Na in their structures. Consequently, basic rocks contain considerable Na but subsequent rocks contain even more, because the plagioclase composition shifted to that of albite and the micas containing Na crystallized at later stages. Potassium, Rb, and Cs are mainly found in micas and K-feldspars, which are late crystallized. These cations occupy the same position in silicate structures; their differentiation therefore arises mainly from the greater abundance of K and the much larger size of Cs ions. The net results of the magmatic differentiation of alkali elements is the existence of maximum amounts of each alkali in the acidic rocks. Because of extreme fractionation, minerals with high concentrations of even the scarce alkalies are found in the rocks formed at very late stages and in pegmatites (e.g., 34% Ca in pollucite).

Sedimentary rocks constitute only 5% or so of the Earth's crust but they have an important bearing on soils because they cover about 75% of the land surface (Pettijohn, 1957) and, therefore, serve as parent materials for much of the soil. An average sedimentary rock includes 80% shale, 12% sandstone, and 8% limestone according to Horstman (1957). The alkali contents of these main sedimentary rocks have been compared in Table 3. Because of their high clay content, the shales contain the largest amount of each alkali and are even enriched in Li, Rb, and Cs relative to the average igneous rock (crust). Residual particles of feldspar and mica along with clay impurities account for most of the alkali content of sandstones and limestones. As sandstones mature and reduce mainly to quartz, their alkali content becomes quite small. Similarly, with limestones the level of alkali decreases as the purity of the carbonate increases.

Because the tendency for clay minerals to adsorb alkali cations from solution increases in the order Li < Na < K < Rb < Cs (see discussion of Table 1), the degree of alkali retention by argillaceous sediments should increase in the same order. Except for Li, the alkali elements seem to be enriched in average shale (Table 3) and various sediments (Heier and Adams, 1964; Welby, 1958) in accordance with these differences in exchange selectivity. Being a weakly held exchangeable cation, Na remains in solution and is carried out to sea by streams if there is adequate water. As a result, sediments contain less Na than an average igneous rock. If water is lacking, as in arid or semiarid regions, saline and sodic conditions can arise from the abundance of dissolved Na. Potassium is adsorbed strongly enough by clays to avoid much loss but even more of the Rb and Cs cations released by rock weathering may be retained by the clay minerals formed in the vicinity. However, the relative enrichment of Cs > Rb > K anticipated with those cations and observed for some shales (Heier and Adams, 1964) is not always evident in sediments (Welby, 1958). Because Li is the least tightly held exchangeable cation, a loss of Li similar to that of Na may be expected during sedimentation. This does not usually occur because Li can occupy octahedral positions in the clay minerals and be strongly held as part of the mineral structure. Specific enrichment of alkalies in sediments because of their retention in the interlayers of contracted layer silicates might be expected as well. However, associations between the alkalies and specific clay minerals are still considered to be uncertain (Chester, 1965).

The average values for the alkali concentrations in soils (Table 3) were based on several sources of data. Considering the range of concentrations that are listed by Bowen (1979) for each of the alkalies and more recently by Kabata-Pendias and Pendias (1984) for Li, Rb, and Cs in a variety of soils, these average values have limited use. In general, however, they reflect the relative effects of pedogenesis on the different alkalies. It is evident that soils have lost considerable Na and lesser amounts of K but have maintained or increased their concentration of Li, Rb, and Cs.

The composition of the stream water is determined by the relative loss of alkali cations during the weathering of rocks, soils, etc. The greatest loss occurs with Na, which tends to stay in solution and build up a high concentration in the sea. Less K is removed by the stream water and even less accumulation of K is observed in the sea because part of the K reaching the sea is adsorbed by the clay minerals in the marine sediments

(Weaver, 1967). The selective retention of K by soil clays accounts for there being more Na than K in the ground water, but this order is reversed in the runoff water because the surface soils contain much more K. The difference in rainfall content of Na and K appears to result from sea spray (Tinker, 1967a).

The levels of Li, Rb, and Cs in stream water and the sea are quite low because the small amounts of these elements that exist in soils, rocks, etc., are strongly retained by the surfaces and structures of the minerals. According to Angino *et al.* (1974), a mean Li level of about 2 $\mu g\ L^{-1}$ is typical for the water supplies of many United States cities, but levels up to 380 $\mu g\ L^{-1}$ have been reported.

# IV. Mineralogy

The bulk of the alkali elements in the Earth's crust occurs in feldspars and micas (Heier and Adams, 1964), but many other minerals contain these elements and various combinations of these minerals can be responsible for the alkalies in particular rocks and soils. Some indication of the different proportions of alkali-bearing minerals encountered in rocks can be obtained from a comparison of average igneous and sedimentary rocks. According to Ahrens (1965), average igneous rocks contain 31% alkali feldspar (1 : 1 orthoclase–albite), 29.2% plagioclase, 12.4% quartz, 12% pyroxene, 4.1% oxides of Fe and Ti, 3.8% biotite, 2.6% olivine, 1.4% muscovite, and 3.5% other minerals. By comparison, sedimentary rocks with 7% feldspar, 38% quartz, 20% mica, 20% carbonate, 9% clay, 3% limonite, and 3% accessory minerals (Jeffries, 1947) have much less of their alkalies in feldspars and especially in pyroxenes, because these minerals are lost by weathering, and have more of their alkalies in micas and clay minerals. Such differences in mineral combinations have significance because the minerals differ in respect to the location and amount of each alkali element they contain. Therefore, specific information about the mineralogy of the alkali elements is pertinent to a description of the sources, forms, and amounts of alkalies in soils. Lists of the numerous minerals involved and histograms showing their alkali content are given by Heier and Adams (1964) and by Heier and Billings (1970). Details regarding many independent alkali minerals and alkali carriers are given by Vlasov (1966a, b).

The mineralogy of Li is distinct from that of the other alkalies because Li is enough different in size (about 20% smaller than Na) that isomorphous substitution with even Na is very limited. Instead, because of their similar size, Li is commonly associated with Mg (radius 0.078 nm) and Fe (radius 0.082 nm) in octahedral positions of the silicate minerals. Lithium forms a more ionic bond with oxygen than Mg or Fe (difference in electronegativity values being greater), but the higher charge and greater abundance of Mg and Fe give them preference for these crystal sites. Thus, Li is found in late olivines, pyroxenes, and amphiboles but in amounts that vary with the magmatic stage at which these minerals have formed. Feldspars do not provide structural sites suitable for Li; therefore, their content of Li is low (generally <5 mg $kg^{-1}$) and seems to be associated with a substitution of Li for Na in plagioclase. On the other hand, Li is commonly found in micas and clay minerals because it is readily accommodated in the octahedral layers of these minerals and the competition from Mg and Fe was less during the formation of these minerals. There are various micas with very little octahedral Li, but others contain enough Li to be called Li micas (Stevens and Schaller, 1942). According to Foster (1960), Li micas can be interpreted as being related to muscovite by Li substitution for Al (2.5 Li per Al) or siderophyllite by Li substitution for $Fe^{2+}$ (2 Li per 1.5 Fe) with the two series joining at lepidolite and with polylithionite as an ideal end member.

Although it is well known that most of the indigenous Li in micas occurs in the octahedral sheets of the mineral, Robert *et al.* (1983) have concluded from hydrothermal experiments with synthetic trioctahedral micas that micas must exist in nature with interlayer Li. The Li in such micas is considered to be anhydrous and not located in the usual K site. The presence of some interlayer Li may also result from an exchange of Li for interlayer K in micas, but the micas would be expanded by the presence of hydrated Li (Rausell-Colom *et al.*, 1965).

In addition to the silicates, many phosphates, borates, and other minerals are known to contain Li. More than 80 minerals have been listed as Li carriers and 25 minerals have been described as independent Li minerals by Vlasov (1966a,b). Included in the latter group are the more common Li minerals: spodumene (inosilicate); lepidolite, polylithionite, and zinnwaldite (phyllosilicates); petalite (tectosilicate); amblygonite (phosphate); and cryolithionite (halide). The Li content of these minerals is only 30 to 120 g $kg^{-1}$ because of the low atomic weight of Li and the inclusion of other cations for mineral stability. Moreover, these minerals occur mainly in pegmatites because adequate levels of Li in the residual magma did not develop until the late stages of igneous rock formation.

According to Vlasov (1966a), sedimentation processes promote a great deal of Li dispersion. This is because much of the Li in primary minerals is released by weathering and is distributed throughout the sediment as part of the clay mineral structures. The average amounts of Li reported by Horstman (1957) for several samples of kaolinite, montmorillonite, and mixed layer clays are 120, 47, and 47 mg $kg^{-1}$, respectively. McLaughlin (1958, 1959) found much less Li in the clay fractions of many kaolinites (2 to 50 mg $kg^{-1}$) but about the same amount in illites (40 mg $kg^{-1}$). Whereas most smectites contain too little Li to be reported in analyses, the clay mineral hectorite and a few clay deposits have been shown to contain up to 5.7 g Li $kg^{-1}$ (Deer *et al.*, 1962; Norton, 1965).

The distribution of Na in minerals is governed not only by its size but by its abundance and the extreme solubility of its salts. Unlike Li, the bulk of the Na in the Earth's crust is located in feldspars. However, some Na occurs in pyroxenes and amphiboles as well. This is possible because Na is enough larger than Li to have the six- or eightfold coordination number required by the pyroxene and feldspar crystal structures, respectively. Potassium is also found in feldspars, but the association between Na and K in these minerals is limited by the differences in their sizes. At high temperatures, solid solutions between Na and K feldspars (e.g., anorthoclase) exist, but a separation of the two feldspars occurs when the temperature drops (Barth, 1969). On the other hand, Na is more like Ca (radius 0.106 nm) in size and is found to be closely associated with Ca in many minerals of igneous rocks. Solid solution series exist between anorthite and albite (Ca and Na feldspars) and between diopside and aegirine (Ca and Na pyroxenes). Consequently, the amounts of Na in feldspars and pyroxenes vary widely and depend upon the magmatic stage involved. The hornblendes also contain variable amounts of Na. Other tectosilicates contributing to the Na content of igneous rocks are the feldspathoid minerals, nephaline and sodalite, which account for much of the Na in alkaline rocks, and the zeolites (e.g., analcime), which are secondary-origin minerals. Micas are also important Na carriers, but the Na in these minerals is located between the layers. The only Na mica, paragonite, is generally considered to be rare, but reports of paragonite in soils (Cook and Rich, 1962) suggest that it may be more widespread than previously thought. Naturally occurring Na vermiculite has also been found (Curtis *et al.*, 1969).

Because pyroxenes, amphiboles, feldspathoids, and some of the plagioclases are very susceptible to weathering, much of the primary minerals containing Na in igneous rocks is lost during sedimentation. The authigenic minerals of sedimentary rocks lack Na because Na is not easily immobilized once it is in solution. Therefore, most of the Na in sedimentary materials occurs in detrital feldspars and micaceous minerals. Because Na is not strongly adsorbed on the surface of clay minerals and does not promote a contraction of silicate layers, the Na content of clay minerals is usually low. Typical values for the Na content of kaolinite, montmorillonite, and illite are 1.2, 1.6, and 4.4 g $kg^{-1}$, respectively (calculated from data in Deer *et al.*, 1962).

The main minerals containing K in igneous rocks are feldspars (microcline, orthoclase), feldspathoids (leucite), and micas (biotite, muscovite). Because most of these minerals are late crystallates, K is particularly abundant in the acidic rocks but the presence of some biotite in more basic rocks gives K a wider distribution. Because of its size, K does not occur with Li and Na in octahedral positions in the structures of pyroxenes and amphiboles, or to much extent with Na in feldspars. The three alkalies do occur together in micas, but Li is in the octahedral layer whereas K and Na are in interlayer positions, and the prevalence of K in micas is much greater. On the other hand, isomorphous substitution of Rb for K is very general and limited only by the scarcity of Rb because these cations have similar sizes. For the same reason, solid solutions between K and Ba feldspars exist to the degree allowed by the scarcity of Ba (radius 0.143 nm). During sedimentation, leucite and biotite are decomposed more readily than K feldspars and muscovite. Thus, much of the K in sediments occurs in residual K feldspar and dioctahedral mica particles. In addition, part of the K released by igneous rock weathering is removed from solution by ion exchange and retained as interlayer cations in contracted layer silicates. As a result, illites can have an average K content of about 60 g $kg^{-1}$, whereas kaolinite and montmorillonite often contain only 2 to 3 g $kg^{-1}$ (Deer *et al.*, 1962).

Rubidium forms no independent minerals. Instead, because it is scarce and very similar to K in size and electronegativity, Rb occurs in most K minerals as a substitute for K. Therefore, there are many Rb carrier minerals but only a few of these are considered principal carriers and have Rb contents up to 45 g $kg^{-1}$. Some of the minerals with the most Rb are microcline, amazonite, muscovite, lepidolite, and zinnwaldite (Vlasov, 1966b). Micas and K-feldspars are the most important Rb-carrying minerals in that most of the Rb in the Earth's crust is located in these minerals. Typical levels of Rb in these minerals, according to the histograms of Heier and Billings (1970), are 300, 600, and 600 mg $kg^{-1}$ for K-feldspar, biotite, and muscovite, respectively. Such data and K/Rb ratios for coexisting K-feldspar and mica show that there is a preference for Rb to be incorporated in micas instead of feldspars (Lange *et al.*, 1966) as expected from the larger size of the Rb ion. However, the bulk of the Rb in igneous rocks occurs in the more prevalent K-feldspars. The distribution of Rb in rocks and minerals is similar to that of K; that is, these alkalies are geochemically coherent, the K/Rb ratios being mainly confined to a range of 160 to 300 (Heier and Billings, 1970). Rubidium is too large to substitute for Li but does occur in Li minerals where substitution for K is possible (e.g., lepidolite). Some substitution of Rb for Na occurs but Ahren's (1965) comparison of plagioclase and K-feldspar from the same granite specimen but with 20 and 300 mg Rb $kg^{-1}$, respectively, shows the greater preference Rb has for K sites. Because Rb ions are not much smaller than Cs (about 10% less), it is not surprising that appreciable amounts of Rb are found in the only independent Cs mineral, pollucite. Also, because of their size similarity, Rb and Tl (radius 0.149 nm) are commonly found together in micas and feldspars even though Tl forms a much more covalent bond with oxygen.

During sedimentation, Rb and K are released together by the weathering of igneous rocks, but relatively less Rb is lost in stream water because Rb is held more strongly by clay minerals. Compared to the Earth's crust, therefore, argillaceous sediments generally have lower K/Rb ratios and more of their Rb in clay minerals. Average values for the Rb content of various clays are reported by Horstman (1957) to be 110, 25, and 140 mg $kg^{-1}$ for kaolinite, montmorillonite, and mixed layer clays, respectively. Most of the clay fractions of kaolinite analyzed by McLaughlin (1959) did not contain detectable amounts of Rb, but one contained 140 mg $kg^{-1}$. The clay fractions of Fithian and Morris illites are reported to contain 700 and 1000 mg Rb $kg^{-1}$, respectively (McLaughlin, 1958).

Most of the Cs in the Earth's crust is located in micas and K-feldspars as a substitute for K. The degree of isomorphous substitution of Cs for K in these minerals, however, is limited by their size difference (the K radius being 19% less than that of Cs) and is less than one-half the Rb substitution for K (Greenwood, 1960) despite the more ionic character of the Cs bond with oxygen. Consequently, the level of Cs in most of these minerals is low and very dependent on differentiation and pegmatite formation. Potassium-feldspars generally contain less than 100 mg Cs $kg^{-1}$ even if they occur in pegmatites, but values up to 3000 mg $kg^{-1}$ have been reported for samples from pegmatites (Heier and Adams, 1964). Because of its size, Cs is like Rb in entering interlayer positions in micas in preference to feldspars, and according to Heier and Billings (1970) muscovite typically contains more Cs than does biotite. The Cs concentrations observed with micas seem to range from 5 to 30 mg $kg^{-1}$ for biotite, 140 to 3100 mg $kg^{-1}$ for muscovite, and can exceed 10 g $kg^{-1}$ in lepidolites (Heier and Adams, 1964; Norton, 1965). Other minerals with high concentrations of Cs include beryl, rhodizite (borate), and avogadrite (halide) with up to 70 g $kg^{-1}$ (Vlasov, 1966a). The only independent Cs mineral is pollucite, which contains 340 g Cs $kg^{-1}$. In sediments, the distribution of Cs appears to be related to the clay minerals present, as would be expected from the specificity with which Cs is held as an exchangeable ion. However, data on Cs concentrations in clays are still very scarce. The average amounts of Cs reported by Horstman (1957) for kaolinite, montmorillonite, and mixed-layer clays are 14, 1, and 17 mg $kg^{-1}$, respectively.

In soils, the alkali-bearing minerals are the same as those in igneous and sedimentary rocks but their relative proportions are usually altered by the soil-forming processes. A discussion of these minerals in soils and their alteration during soil development has been presented by Barshad (1964) and Jackson (1964). Of importance here is the fact that feldspars and micas continue to be the most abundant alkali-bearing primary minerals in soils, but much of the alkali elements in soils is associated with clay minerals, particularly illite.

## V. Soil Content

### A. Range of Content and Geographical Distribution

The alkali content of soils is a function of the parent material and the many chemical, physical, and biological forces imposed on it. Consequently, as shown in Figure 1, the range in content of each alkali element

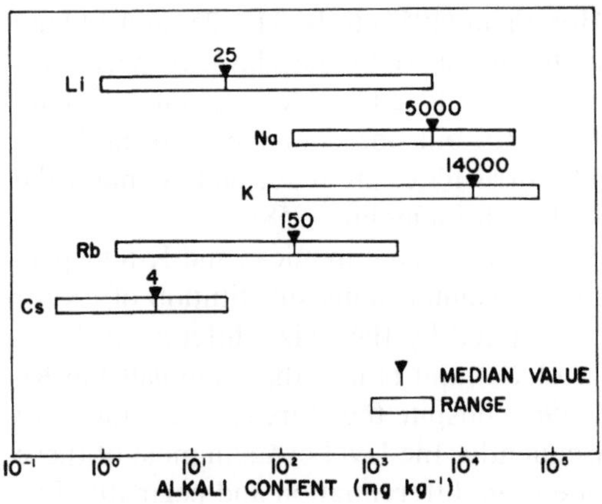


**Figure 1.** Medians and ranges of alkali concentrations in soils. Median values from Bowen, (1979); ranges based on data from Bowen (1979), Heier and Adams (1964), and Marbut (1935).

is considerable. The significance of these variations in the amounts of alkalies, of course, depends on the general level and role of each element in soils. The fact that K is a major plant nutrient makes its variations particularly important. The content of K in soils is not only more variable but also greater than that of other plant nutrients. In the plow layer of cropland in the United States, the K content is reported to be 5.8 times that of nitrogen and 13.4 times that of phosphorus. Median values are given in Figure 1 to emphasize the relative amounts of the alkalies that usually occur in soils. These median values are very similar to the average values listed in Table 3.

Despite the existence of a range in alkali contents, the distribution of alkalies in soils follows a fairly definite geomorphological pattern that reflects both the differential weathering of mica, feldspar, pyroxene, and amphibole in the parent materials and the subsequent association of the released alkalies with clay minerals. In general, the differential weathering follows the reverse order from that which the alkali-bearing minerals form when molten magma cools. That is, those minerals which formed first tend to weather most readily. The importance of any primary mineral as a factor influencing the alkali content of soils will decrease as the soil ages. This is evident in the old, highly weathered, sandy soils of the southeastern states, which have relatively low alkali contents even though they formed from rocks that were once quite rich in mica and feldspar. On the other hand, the younger, less weathered, shale-derived, till soils of the Midwest still contain appreciable quantities of mica and feldspar and have relatively large amounts of the alkali elements. However, even in these soils, weathering has been intense enough to remove considerable amounts of the original alkali elements. This is especially so in the case of Na, which is not strongly sorbed by clay minerals and is easily lost by leaching. Soils with high Na contents are more likely found in arid regions, where leaching has been minimal.

The geographical distribution of the alkali content of soils tends to vary inversely with mean annual temperature and precipitation. This generalization is evident from the numerous chemical analyses of soils compiled by such workers as Coffey (1912) and Marbut (1935) and is illustrated by Reitemeier's (1957) observation that the K content of United States soils generally increases from east to west (i.e., in the direction of less severe weathering) and from south to north in the eastern half of the United States. Nevertheless, many exceptions to this trend in alkali distribution exist because of differences in parent material, age, drainage, etc. For example, the Durham and Cecil soils of the Piedmont Plateau show the differences in content that can arise from drainage. Both soils have developed from similar igneous rock parent material, but the poorly drained Durham soils contain about 40 g K $kg^{-1}$, whereas the well-drained Cecil soils usually contain less than half that amount (Marbut, 1935). Additional information on the geographical distribution of alkalies in soil is given by Jackson (1964), Mitchell (1964), and Vinogradov (1959).

Some of the information that has been published about the amounts and distribution of alkali elements in soils has been related to soil subdivisions or regions. Vinogradov (1959), for instance, presented his data for Li, Rb, and Cs in terms of Podzolic, Grey Forest, Chernozem, Chestnut, Serozem, and Red soils. More recently, differences among the soils of temperate and boreal, arid and semiarid, and tropical humid regions (Aubert and Pinta, 1977) and the soils of the United States or different countries (Kabata-Pendias and Pendias, 1984) have been described. Detailed relationships between the alkali content of soils and current soil classification schemes, however, are still lacking. On the other hand, the following generalization can be made (D.A. Laird, Iowa State University, Ames, IA, personal communication, 1985) at the order level of soil taxonomy (Soil Survey Staff, 1975). The alkali content of Entisols and Inceptisols varies with their parent material, whereas Mollisols and Alfisols have relatively high levels of alkalies because they contain considerable amounts of unweathered minerals. Vertisols have a variable alkali content that depends on local conditions. Histosols, Ultisols, Spodisols, and Oxisols reflect the effects of prolonged and intense leaching environments by having low alkali levels. Of the ten soil orders, Aridosols have the highest content of the alkali elements. In addition, the strong influence of the soil moisture regime on the alkali content of soils must be accounted for. Generally, the alkali levels are highest in Aridic soils, lower in Ustic and Xeric soils, and even lower in Udic soils. Variable amounts of the alkalies occur in Aquic moisture regimes. The presence of a natric horizon (an argillic horizon having a sodium absorption ratio $\geq 13$ and typically columnar structure) or a salic horizon (one that is at least 15 cm thick and contains >2% secondary enrichment of salts that are more soluble than gypsum in cold water) is an important criterion of soil classification.

**B. Particle Size**

Alkali cations occur in all the particle sizes of soils but are not always in the same minerals. In sand and silt particles, most of these cations occur in minerals (e.g., mica and feldspar) that have persisted from the parent material. In the clay fraction, the alkalies are mainly associated with clay minerals. Consequently, the abundance of various alkali-bearing minerals in different size fractions of the soil has received much attention. Typical observations are given by Black (1968) in a summary of reports on detailed K mineral analyses of seven German soils. On the average, soils with K levels of 23 g $kg^{-1}$ contained 27% (by weight) K minerals (15.5% feldspar and 11.3% micaceous minerals). Of the feldspar, 93% was in the sand and silt and 7% was in the clay. Of the micaceous minerals, 24% was in the sand and silt and 76% was in the clay. The portion in the clay was probably illite.

The alkali cations in soils are not uniformly distributed among the different particle sizes. Bolton (1966) found the clay fraction of British soils contained less Na, but as much or more K, than the silt and fine sand fractions. Such differences result from the persistence of coarse Na-feldspar particles and the retention of much more K than Na by clay minerals.

Although differences in parent material and weathering intensity preclude a definite relationship between particle size and K content, it seems that most often the K content is highest in the clay particles. Failyer *et al.* (1908), in their study of 27 surface soils from different regions of the United States found that K content usually increased as the particle size decreased. In a summary of this work Buckman and Brady (1969) report the highest K contents (42 g $kg^{-1}$) were in the clay particles of arid region soils, whereas the lowest K contents (3.1 g $kg^{-1}$) were in the sand particles of coastal plain soils. The greatest variation in K content with particle size was also in coastal plain soils, where the K content of sand, silt, and clay particles averaged 3.1, 11.0, and 13.4 g $kg^{-1}$, respectively. On the other hand, Salmon (1964) analyzed 22 surface soils and eight subsoils of Rhodesia (now Zimbabwe) and found the trend for the K content–particle-size relationship to be generally the reverse; that is, the K content tended to decrease with particle size. Such differences in results can be accounted for by the relative proportions of the soil minerals in the various size fractions. By separating soil fractions with narrow particle size limits, Liu (1970) found the K content of the fractions decreased as the particle size increased or decreased from the 0.2- to 2-μm range and related these changes to an increase in smectite concentration in the finer fractions and an increase in quartz in the coarser fractions.

Information regarding Li, Rb, and Cs contents in different sizes of particles is meager. Often, in the case of Rb and Cs, their content in the sand and silt particles is so low they cannot even be detected. Aidinyan (1959) analyzed Red and Yellow, Chernozem, Brown Forest, and Alluvial soils of Russia and found Li, Rb, and Cs to increase in content quite consistently as particle size decreased. In the case of a Chernozem surface soil, the contents of Li, Rb, and Cs in the 0.2-μm clay were 100, 500, and 50 mg $kg^{-1}$, respectively, whereas the corresponding values for the whole soil were 20, 150, and 10 mg $kg^{-1}$. In a study involving four Scottish soils Mitchell (1964) reported that Li and Rb content in the clay particles was usually five to ten times that in the sand particles.

In regard to alkali content–particle-size relationships, it is well to note that, even though one particle size may contain less alkali than another, the lower content particles may still account for the majority of the alkali if they comprise the bulk of the soil. In the study of Failyer *et al.* (1908), however, it usually made no difference whether the K contents were calculated on a whole-soil or a size-fraction basis.

## C. Profile Distribution

There is no distinct trend to the changes in alkali content that occur with depth in the profile. In soils of initially uniform parent material, the alkali content may remain about the same or increase with depth, depending upon the extent of chemical weathering and leaching. Sometimes, surface soils are found to have a higher alkali content than subsoils. Generally, such a situation reflects the action of plants in transporting alkali cations to the surface. Burrowing animals, such as earthworms, termites, and rodents, can also contribute to a nonuniform distribution of alkalies by moving soil from one layer to another, but their net effect is more likely one of homogenizing the soil. Under irrigated conditions in arid regions upward capillarity and evaporation may concentrate alkali cations (typically Na) at the soil surface. In Solonetz soils, weathering has not been intense enough to move much Na out of the profile. Instead, the Na is concentrated in a natric horizon.

In a detailed study of four soils (igneous rock parent material) from Wyoming, Colorado, Missouri, and North Carolina, Short (1961) observed that the Na and K contents remained about the same or increased with depth. McLaughlin (1955), in a study of nine New Zealand soils (arkosic parent material), observed slight increases in the concentrations of Li and Na in some surface horizons, even though there had been some leaching of both cations from the profile. Subsequent work on the distribution of Li in the profiles of various azonal, zonal, interzonal, and Podzolic soils of New Zealand (Wells and Whitten, 1972) has brought out the modifying effects of leaching and parent material but the distribution was generally uniform in the profiles. Potassium and Rb show two distinct characteristics, an accumulation in the surface by plants and a concentration deeper in the profile because of sorption by clay. Swaine and Mitchell (1960), in a study of eight representative Scottish soils (igneous rock parent material) which included Podzols, Brown Forests, and Gleys, found that within the limits of detection Li and Rb exhibited no great variation with depth. Butler

(1954), however, in a study of six British Gleys, observed a fairly regular increase with depth for Li and Rb. Moreover, the increase was independent of drainage. Burridge and Ahn (1965), in a study of 19 Ghanaian forest soils (phyllite, granite, and sandstone parent materials), observed some tendency for Li, K, and Rb to increase with depth, but there was no well-defined trend for Na. Aidinyan (1959) reported that Red and Yellow, Brown Forest, Chernozem, and Alluvial soils in Russia had surface horizons enriched with Li, Rb, and Cs because of root transport from lower horizons. Vinogradov's (1959) data show that Li, Na, K, and Rb contents of Russian Podzols and Chernozems tended to be more or less uniform with depth in the profile.

In many of the aforementioned cases specific alkali cation contents varied by less than a factor of two. Relative to the distribution of many other chemicals in the profile, such variations are minor. Marbut (1935) has presented a pictograph of the changes in chemical content with depth to 5 ft for 12 representative United States soils. A striking feature of this pictograph is that the degree of variation of alkali cation content with depth is much less than that for silica, alumina, iron oxide, alkaline earths, and organic matter.

### D. Fertilizer, Organic Matter, and Precipitation Effects

Despite the fact that agricultural soils may contain 3 to 30 g total K $kg^{-1}$, modern cropping practices generally include the application of K fertilizer. Often, release of K from mineral forms is too slow to support optimum crop yields. In 1984, harvested cropland in the United States received an average addition of 32 kg K $ha^{-1}$; in 1965, by comparison, the addition was 17 kg $ha^{-1}$ (Hargett and Berry, 1985). Although small in comparison to the total amount of K in soils, these fertilizer additions represent readily available sources and can be quite important to the crop. The K additions are most commonly in the form of KCl.

A compilation of the elemental compositions of fertilizers for the major plant nutrients has been prepared by Swaine (1962). It is evident that fertilizers supplying other major nutrients can contain appreciable amounts of Na and K (up to 25% Na and 35% K in some nitrogen fertilizers). The amounts of trace alkalies in fertilizers are quite low—$<30$ mg Li, $<150$ mg Rb, and $<1$ mg Cs $kg^{-1}$.

As shown by Bowen (1979), all the alkali elements occur in plant and animal tissues. In fact, it was their presence in plants that provided some of the first indications that trace alkalies existed in soils. As long as the tissues are living, little loss of these alkalies to soils can occur. However, it may be concluded from K experiments by Grimes (1966) that most of the alkalies in crop residues (including dead roots) diffuse from the residues very quickly when these residues are incorporated with soil. Moreover, it has been observed that soil organic matter has a low affinity for alkali cations (McGeorge, 1931; Baver and Hall, 1937). As a result, the alkali content of soil organic matter tends to be quite low. Most methods of extracting soil organic matter (Mortensen and Himes, 1964) preclude a determination of the alkalies that are associated with the organic fraction. However, the presence of only 0.08% K in a representative peat surface soil with 80% organic matter (Buckman and Brady, 1969) shows there cannot be much involved. Some enrichment of Li, Rb, and Cs in humic horizons of Russian soils has been reported (Aidinyan, 1959), but Mitchell (1964) found no enrichment of Li or Rb in organic-rich surface horizons of Scottish soils.

Because farm manure is often applied to soils at rates of 22 to 44 Mg $ha^{-1}$, the manure can serve as an important organic K source in some soils, even though the K content of representative farm manure is only about 4 g $kg^{-1}$ (Brady, 1984). Likewise, sewage sludge with a representative K content of 2.4 g $kg^{-1}$ may also serve as a K source to some soils. Typically, the Na content of farm manure is severalfold less than that of K (US Department of Agriculture, 1979), whereas in sewage sludge the Na content may be comparable to that of K (Kirkham, 1979). At least 25% of the sewage sludge in the United States is currently land applied, and the percentage is expected to increase as alternative disposal means become more expensive and less attractive.

The contents of alkali elements in precipitation originate mainly in particulate and dust materials suspended in the atmosphere. The content of Na, especially, is influenced by sea spray. According to Tinker (1967a), about 17 kg Na $ha^{-1}$ is deposited annually by precipitation in areas of Britain that are not near the sea. In the United States, the amounts of Na deposited are often lower, although some higher amounts have been reported (Tabatabai, 1983; Sharpley *et al.*, 1985). Generally, annual amounts of K received in precipitation are $<5$ kg $ha^{-1}$.

### E. Artificial Isotopes

Nuclear explosions and the operation of atomic reactors produce fission products that include $^{137}Cs$ ($t_{1/2}$ = 30 years) and $^{86}Rb$ ($t_{1/2}$ = 18.6 days). Because of its short half-life, any $^{86}Rb$ reaching soil has a rather fleeting existence. Consequently, only the $^{137}Cs$ poses a potential long-term problem as a soil contaminant. In general, this nuclide remains on or near the soil surface, usually in the upper 5 to 7 cm (Schultz *et al.*, 1960; Ritchie and McHenry, 1973). However, on sandy soils deeper penetration of the $^{137}Cs$ can occur with leaching. Mechanical disturbance of the soil, such as plowing, may also incorporate $^{137}Cs$ at lower levels. Probably the most extensive recent work in the United States involving fallout $^{137}Cs$ in soils has been conducted by McHenry and associates (e.g., McHenry and Bubenzer, 1985). Their interest stems from use of $^{137}Cs$ measurements to assess erosion losses and sediment deposition. Based on his data, J.R. McHenry

(USDA, ARS, Durant, OK, personal communication, 1985) estimates the $^{137}Cs$ activity of United States soils to range approximately from $2 \times 10^9$ to $4 \times 10^9$ Bq km$^{-2}$ land surface (0.6 to 1.3 mg $^{137}Cs$ km$^{-2}$). The values have been adjusted to 1980, and the higher value would be representative of fairly undisturbed, noneroded areas in the central United States. Fallout $^{137}Cs$ levels in the atmosphere reached a peak in 1963–65 and, with the cessation of aboveground nuclear weapons testing, have continued to decline since then. Even during the period of peak fallout, primary concern centered around the contamination of vegetation rather than soil.

The disposal of $^{137}Cs$ and other radioactive wastes from nuclear installations continues to be a major problem, even though many methods of waste management have been investigated. The concept of using ion-exchange minerals for this purpose has been reviewed by Robinson (1962) and has led to the early use of clays in waste treatments involving Cs removal. At Oak Ridge National Laboratory, shale containing illite was used to sorb the Cs in waste effluents (Tamura and Jacobs, 1960; Lomenick *et al.*, 1967). Laboratory experiments with heated illite have shown that various heat treatments should increase the Cs sorption capacity of the shale (Smith and Scott, 1974). However, the use of clays and other methods of Cs waste management have undergone a variety of changes in recent times (Lee and Tank, 1985).

# VI. Soil Forms

The alkali elements in soils are not easily characterized in terms of different forms because their state throughout the soil is essentially ionic. Their presence as ions in the solution, on the exchange complex, and in the living matter of soils is well known, but it should be kept in mind that the largest part of the various alkalies in soils occurs as an ionic constituent of soil minerals as well. Even so, the concept of alkali forms is deeply ingrained in the literature on these elements in soils. In general, the forms of the alkalies in soils are simply portions of the cations that are delineated by their location in the soil or their extractability by specified chemical or biological methods. These distinctions are not independent; indeed, the extractability of the cations is usually related to their location. Moreover, sharp distinctions between portions of the cations are lacking with these methods of classification. Nevertheless, it is evident from the reviews of potassium forms in soils (Reitemeier, 1951; Wiklander, 1954; Duthion, 1968; Mielniczuk, 1979; Metson, 1980) that some subdivision of the total cations can be quite useful.

As our knowledge of alkali behavior in soils increases, the significance of making specific subdivisions in the cations has become more evident but the terminology used with the alkali forms needs to keep pace to avoid confusion. At present, connotations regarding plant availability are commonly assigned to chemically defined forms of K. For instance, Duthion (1968) states that the distinction among exchangeable, fixed, and definitively fixed forms of soil K rests on their availability to plants. This situation stems from the extensive search that has been made for chemical methods of assessing the plant availability of K in soils. However, as pointed out by Cooke (1967), applying the term "available" to the fraction of a plant nutrient in soil that has been extracted by a chemical reagent can be misleading. Also, the practice of equating alkali forms that have been delineated by different criteria can restrict the development of concepts about the individual forms. For instance, the behavior of most of the exchangeable K in soils is often ignored because it is not utilized by plants.

Most of our information about the forms of alkali elements in soils has been obtained with Na and K because the small amounts of Li, Rb, and Cs involved have created analytical difficulties in the past. In addition, interest in the indigenous forms of the trace alkalies in soils has been slight except for a few instances where the effects of Li forms have attracted attention. However, experiments with minerals that contain more of these alkalies, others involving additions of the alkalies to soils or clays, and some that rely on the enhanced sensitivity of current analytical methods are now providing direct and indirect information about the indigenous forms of the trace alkalies in soils. In this discussion, the general situation regarding alkali forms in soils is developed in terms of this combination of information but pertains primarily to K.

A variety of criteria and experimental conditions has been used to delineate portions of the alkalies in soils, but each portion of the alkali cations that is segregated by a defined procedure can be considered to be a form of the alkalies. That is, the soil K extracted by boiling $HNO_3$ or any of the methods reviewed by Grimme and Nemeth (1978) and Quemener (1979) constitutes a form of K in soils. By the same token, no reference to alkali forms can have much meaning unless the basis for the delineation of part of the cations is clearly defined. The possibilities of doing so are discussed in terms of the commonly referred to categories of soluble, exchangeable, nonexchangeable, and fixed cations, and the relevance of cation locations in soils. In terms of location, the alkali cations can be classified as occurring in the solution or the solid portions of the soil. Cations in the latter location can be further divided into those located in living tissue, those constituting part of mineral structures, and those associated with exposed surfaces of organic and inorganic solids.

## A. Soluble

The presence of Na and K in soil solutions is well established, but it may be assumed that the other alkalies occur in these solutions as well because all the alkali elements are found in various parts of the hydrosphere and in plants grown on soils (Bowen, 1979). This means that a soluble form (as defined by Reitemeier, 1951) of each alkali element probably exists in most soils. According to this definition, only the quantity of alkali cations existing at any one time dissolved in water of a soil under field moisture conditions and relatively unbound by cation-exchange forces is categorized as soluble. Unfortunately, the identification of this part of the alkali elements in soils is not easily achieved by name or practice. The term "soluble" implies a capability of being dissolved and is sometimes used in this sense to denote alkali fractions that have been solubilized by soil treatments. Also, adequate means of characterizing just the cations in solution have been lacking until recently. The advent of in situ determinations of the dissolved alkalies with ion-selective microelectrodes (Xu and Liu, 1982) should alter this situation and focus more attention on this particular portion of the alkalies in soils. Less confusion over the terminology applied to this alkali form is also needed. In particular, it should be noted that references to water-soluble alkalies have a much broader connotation.

All the soil minerals have some degree of solubility in water and the alkali cations on exchange surfaces are subject to hydrolysis. Consequently, varying amounts of the alkali cations associated with the soil solids as well as those in the soil solution can be included in water-soluble alkali determinations. This dissolution effect of water on alkali-bearing minerals has been verified with rocks, feldspars, and micas and is considered to be a major factor in the chemical weathering of silicate minerals (Loughnan, 1969). Even an increase in soil moisture or an addition of enough water for soil–water extracts can alter the amount of K and Na in solution (F.W. Parker, 1921; Eaton and Sokoloff, 1935; Reitemeier, 1946; Bowen and Cawse, 1965). If much larger volumes of water are used with soils, as in percolation experiments, the amounts of water-soluble K can far exceed the $NH_4$-exchangeable K (Hood *et al.*, 1956; Matthews and Smith, 1957). Under field conditions, the levels of water-soluble K can result in annual losses of K by leaching that have been estimated to be as much as 35 kg $ha^{-1}$ for cultivated humid-region silt loam soils (Brady, 1984).

Determinations of the concentration of indigenous Li, Rb, and Cs in soil solutions are usually precluded by the small sample volumes and low concentrations involved. However, on the basis that river waters represent the average composition of the soil solutions in the region that they drain (Livingstone, 1963), the stream water data in Table 3 should be indicative of the average concentrations of these alkalies in soil solutions. Because none of these concentrations exceeds 3 $\mu g\ L^{-1}$, the dissolved form of these alkalies usually attracts little attention. An exception to this exists in those areas where Li in irrigation waters can produce toxic conditions for plants. Bradford (1963) has found some sources of irrigation water can contain up to 0.5 mg $L^{-1}$ and has identified Li toxicity in field-grown citrus receiving irrigation water with only 0.06 to 0.10 mg $L^{-1}$. A study of plant tolerance to Li in solution (Bingham *et al.*, 1964), however, has led to the conclusion that Li levels of 0.5 mg $L^{-1}$ in irrigation water do not directly suggest toxicity is going to occur. Because Ca has been shown to inhibit competitively the absorption of Li by plants (Epstein, 1960), it is probably the relative levels of these cations in solution that need to be given attention. Even less information is available about the amounts of water-soluble forms of the trace alkalies in soils. Working with soil samples from different depths in nine profiles of Sierozem soils that had a mean total Li content of 33.3 mg $kg^{-1}$, however, Shukla and Prasad (1973) found the hot-water-soluble Li contents varied from 0.35 to 6.90 mg $kg^{-1}$ and averaged 0.87 mg $kg^{-1}$.

By far the highest concentration of soluble alkali cations encountered in soils is that of Na in the salt-affected soils that are classified as saline–sodic (Bresler *et al.*, 1982). To identify this condition, determinations of equivalent conductivity (EC) and sodium absorption ratio ($SAR = [Na]/[(Ca + Mg)/2]^{1/2}$, where Na, Ca, and Mg refer to the concentrations of the soluble cations expressed in meq $L^{-1}$) are carried out with saturation extracts. That is, both the soluble Na level and the relative proportions of the major soluble cations have significance in these soils. An EC value of 4 mmho $cm^{-1}$ divides the saline and nonsaline soils, and an SAR value of 15 divides the sodic and nonsodic soils. In soils without saline or sodic conditions, the importance of soluble Na emanates from the value of Na as a plant nutrient and the relationships that exist between Na and K (Tinker, 1967a). With these soils, therefore, measurements of Na activity ratios ($AR = a_{Na}/a_{Ca}^{1/2}$, where $a_{Na}$ and $a_{Ca}$ refer to ion activities in the soil solution) merit attention (Bolton, 1971). By comparing published values of cation ratios that have been determined for K and Na in soil solutions, Tinker (1967a) concluded that there may be seven or more times as much Na as K in the solution of saline and normal soils.

Whereas soluble Na usually accounts for much of the bases in salt-affected soils, considerable amounts of soluble K are occasionally found. By means of 1 : 5 soil–water extracts, some of these soils have been shown to have 2.64 to 47.22 cmol Na and 0.09 to 0.82 cmol K $kg^{-1}$ in solution (Kelley, 1951). On the other hand, normal soils do not contain appreciable amounts of soluble K unless soluble fertilizers or irrigation water containing K has been added. Nevertheless, the importance assigned to the soil solution as a medium for plant growth has perpetuated an interest in the soluble form of K in all soils. The degree and scope of this interest can be seen in a paper by Moss (1969), where much of the literature pertaining to K in soil solutions has been reviewed. For some time, much effort was devoted to the methods and difficulties involved in the determination of soluble K, but the practice of determining the soluble and exchangeable K in soils together evolved as interest in the exchangeable form increased. Today, however, attention has shifted once more to the concentration of K in solution because such information is needed for the intensity term of quantity–intensity relationships and for ion-movement studies. By analyzing the soil solution or dilute salt solutions that have been equilibrated with the soil, values for K activity ratios or K potentials are being determined. This subject and the applicability of such determinations have been reviewed by Beckett (1971).

Compared to the exchangeable K in soils and the K requirements of crops, the amount of soluble K in soils is very small. Still, the range of K concentrations reported for soil solutions is 0.4 to 400 mg $L^{-1}$ (Asher and Ozanne, 1967) and the mode is in the order of 4 mg $L^{-1}$ for many soils (Barber *et al.*, 1963). Thus, the soil solution may often have a K concentration that exceeds the 1 to 4 mg $L^{-1}$ minimum required in nutrient solutions for the maximum growth of many plants (Asher and Ozanne, 1967) and especially the 0.01 mg $L^{-1}$ K level that permitted good growth of barley in a nutrient medium (Williams, 1961). For nitrifying bacteria in a well-stirred nutrient solution, a K concentration of no more than 0.4 mg $L^{-1}$ can be adequate (Welch and Scott, 1959). On the other hand, uncertainties exist as to whether adequate levels of K can be maintained in the soil solution and especially at the root surfaces when soils are cropped. Marked decreases in the mean concentration of K in soil solutions can occur during the cropping period. However, the degree of change in soluble K and its significance to the maintenance of a supply of K at the root surfaces depends upon the K-buffering capacity of the soil, the rate with which K associated with soil solids enters the solution, and the movement of K in solution to the roots. Details regarding the determination and significance of K-buffering capacity values are given by Addiscott and Talibudeen (1969) and the concepts and literature pertaining to the movement of K to plant roots have been reviewed recently (Mengel and Kirby, 1980). The kinetics of K transfers between the solid and solution phases of soils has also received considerable attention but an integration of the various facets of these transfers is still lacking.

One aspect about the soluble alkalies in soils that needs to be emphasized is the fact that small amounts of the fixable cations (K, Rb, Cs) in solution can block the exchange of interlayer cations in micaceous minerals (Hanway, 1954). The significance of this blocking effect has been established with several experiments involving the chemical extraction and plant availability of interlayer K or $NH_4$ in soil and mineral samples (Hanway and Scott, 1956; Hanway *et al.*, 1957; Bassett, 1959; Welch and Scott, 1960, 1961; Rausell-Colom *et al.*, 1965; Scott and Smith, 1966) and is illustrated in Figure 2. The biotite curve describes the usual observation that progressively more K in the mica can be replaced by Na if the concentration of K in the solution phase is reduced. With this mica in a neutral 1*M* NaCl solution, the dissolved K level must be kept below 7 mg $L^{-1}$ for all the K to be exchangeable. Even lower levels of 0.1 mg $L^{-1}$ dissolved K are required for only 17 and 60% K exchange by muscovite and illite,

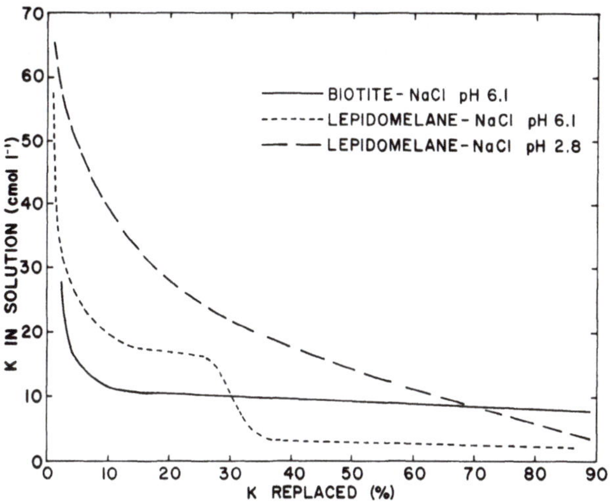


**Figure 2.** Relationship between the equilbrium levels of K in $1M$ NaCl solutions and the degree of K release from 10- to 20- μm samples of biotite and lepidomelane.

respectively (Scott and Smith, 1966), and 0.6 mg $L^{-1}$ for an exchange of 1.7% of the K in a soil (Liu, 1970). Moreover, it is evident from the lepidomelane curves in Figure 2 that the blocking effects of dissolved K can be altered by other factors. In a $1M$ NaCl solution at pH 6.1 the lepidomelane exhibited two distinctly different minimum levels of K selectivity. These results show the mica sample contained a mixture of different particles or particles with different layers. In either event, an increase in the acidity of the NaCl solution reduced the blocking effect of K and obliterated all evidence of the dual character of the mica.

The blocking effects of the fixable alkalies are often described in terms of the concentration of just these cations in solution. Instead, it is actually the fixable/replacing cation ratio in the solution that controls the exchange of interlayer cations (Hanway *et al.*, 1957). This blocking effect is comparable to the mass-action effect of dissolved K on the exchange of K on exposed soil surfaces, but any fixable cation can block the exchange of interlayer K and much larger selectivity coefficients are involved (e.g., $k^{K}_{Na} = 10^6$ for 15% K-depleted muscovite). Because the concentration of replacing cations in natural soil solutions is generally low, the soluble K level must be very low to achieve the K/replacing cation ratio needed for interlayer K release. However, plants must cause this degree of soluble K depletion, at least in zones of the soil, because appreciable amounts of interlayer K in micas and soils can be removed by intensive cropping (Welch and Scott, 1961; Verma, 1963). On the other hand, increasing the level of soluble K (or $NH_4$) in soils by adding fertilizers or plant residues can keep the plants from tapping the vast supply of interlayer K in soils.

The question then arises as to whether the soluble K level can be adequate for plant growth but low enough to permit the release and utilization of interlayer K in a particular soil. The rate with which the K moves to the plant root in this soil is involved in this question because plants can tolerate low levels of K in the solution as long as the level is maintained at the root surface. According to Nye (1972), the concentration of K at the root surface is considerably below that of the bulk solution when the solution level approaches $10^{-4}M$. Under these conditions there would probably be little release of interlayer K unless the associated level of replacing cations were higher than expected. Moreover, the diffusion coefficient values of $2 \times 10^{-5}$, $10^{-6}$, $10^{-8}$, and $<10^{-20}$ $cm^2\ s^{-1}$ reported by Nye for K in aqueous solution, moist soil, dry soil, and unexpanded illite, respectively, show the mobility of the interlayer K could be a limiting factor even if it were released.

Recent advances in ion-selective electrode (ISE) technology are providing many new opportunities for research on the alkalies in soils. In a general review of ISEs, Talibudeen and Page (1983) have described the electrodes that are available for Na and K determinations and the relatively few soil experiments that involved their use. Glass membrane Na and K ISEs have been available since the mid-1950s and have a rugged character that makes them well suited for soil applications, but their selectivity is often inadequate (e.g., $k_{K.NH_4} = 0.1$). Electrodes based on neutral carriers have improved this situation for Na and K (e.g., $k_{K.Na} = 2 \times 10^{-5}$ for valinomycin-based K electrodes versus $7 \times 10^{-2}$ for glass), and the development of new solid state and liquid membrane electrodes are extending the applications of ISEs to Li, Rb, and Cs (Arnold and Meyerhoff, 1984).

The characteristics, performance, applications, and even the preparation of K ISEs have been studied extensively by Farrell (1979, 1985). As with all the alkalies, interferences from other alkalies, $NH_4$, and pH can be a problem when they are used with soil systems. Tolerance for Na has been shifted to higher concentrations by the use of valinomycin-based K ISEs and modified procedures have made it possible to eliminate much of the problem with $NH_4$ in soil extracts. The commonly used reference electrodes with saturated KCl salt bridges cannot be used with K ISEs. The suitability of some of the suggested alternatives for the salt bridge are questionable but saturated LiOAc seems to work well with K solutions. If the ISE determinations of soluble K are carried out with soil suspensions, soil pastes, and field-moist soil samples, however, difficulties with junction potentials and $NH_4$ interferences can be expected. Nevertheless, in situ determinations of K in soils are now being made with microelectrodes (Xu and Liu, 1982; Xuan, 1982).

### B. Exchangeable

Concepts regarding exchangeable alkalies in soils and procedures for their determination stem from the vast amount of work that has been done on

K. Therefore, the general aspects of this discussion are developed in terms of K.

The difficulties involved in defining and determining exchangeable K in soils as expressed by Reitemeier (1951) are as valid today as they were 35 years ago. They are (1) the absence of a sharp distinction between soluble and exchangeable fractions, (2) the existence in some soils of exchangeable K that is not easily or immediately extractable by the usual reagents, and (3) the dissolution of mineral K by the exchange extractant. Also, the bulk of the soil K is still regarded as being nonexchangeable because the methods of determining exchangeable K are geared to the extraction of portions of the soil K that may be equated with its plant availability. Generally, less than 1% of the total K in soils is considered to be exchangeable (Buckman and Brady, 1969). By studying the factors that govern the exchangeability of K in minerals and applying the acquired information to soils, however, it has been shown that up to 47% of the K in some soils is exchangeable (Smith *et al.*, 1968). It is obvious that our concepts about this form of soil K need to be clarified.

In terms of the ion-exchange principles discussed by Helfferich (1962) and the definition Black (1968) has given for exchangeable bases in soils, all the soil K that can be stoichiometrically interchanged with other cations can be included in the exchangeable fraction. Potassium that is sorbed on the exposed surfaces of the soil constitutes part of this exchangeable fraction, but much of the interlayer K in contracted layer silicates is now known to be exchangeable as well (Scott and Smith, 1966). For a variety of reasons, however, all this exchangeable K is not usually determined. Mainly, the degree of exchange depends on the extraction period and the extent to which the replacing cation and procedure limit the blocking effects of fixable cations in solution. With a replacing cation that tends to expand the layer silicates and a method that minimizes the effects of fixable cations (e.g., one that removes these cations by leaching, precipitation, etc.), a large part of the interlayer K can be included in the exchangeable K determination. The rate and degree of this exchange is illustrated in Figure 3 with the results obtained with samples of the <2-μm fraction of Marshall subsoil in a NaCl–sodium tetraphenyl boron ($NaBPh_4$) extracting solution. As much as 25.5 cmol K $kg^{-1}$ (54% of the total K) was replaced by Na under these conditions, whereas the $NH_4$-exchangeable K in this clay was only 2.3 cmol $kg^{-1}$. Even with these small particles, the maximum exchange of K by Na required a period of 3 weeks. On the other hand, it is evident that part of the K in soils is truly nonexchangeable because it is located in physically inaccessible sites in the structures of feldspars. The amount of feldspar K in this clay was determined with a selective dissolution method (Liu, 1970). In addition, the results in Figure 3 illustrate the fact that some of the K in small mica particles (<5 μm) resists exchange by conditions that promote complete exchange in coarser particles (Scott, 1968). In this soil, particle size accounted for 30% of the total K being not exchanged. Various explanations for this particle size effect have been

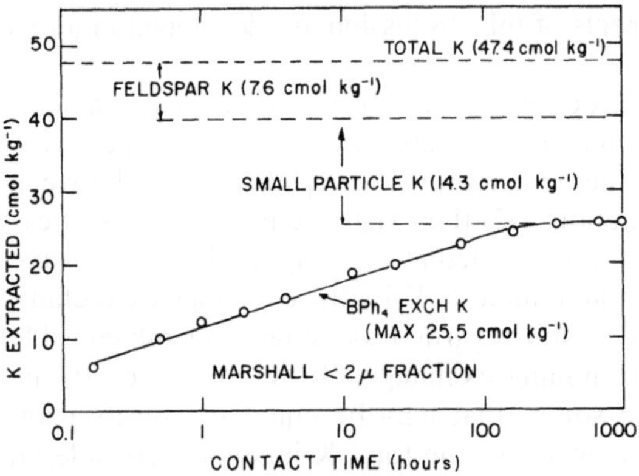


**Figure 3.** Potassium extracted from the <2-μm size fraction of the C horizon of Marshall soil by different contact periods with NaCl–$NaBPh_4$ solutions at 25°C.

proposed (Norrish, 1972), but the possibility has not been ruled out that some or all of this portion of the soil K may be exchangeable under the right conditions.

If the convention of using $NH_4$ salts to displace exchangeable K (Black, 1968) is employed, the interlayer K in contracted layer silicates is largely excluded. Therefore, $NH_4$-exchangeable K represents a minimum limit to the amount of soil K that should be classified as the exchangeable form. In general, $NH_4$ replaces more of the soil K than other replacing cations when a procedure that does not promote the exchange of K in contracted layer silicates (equilibrium extraction, small extractant volume per increment of soil, etc.) is involved (Scott and Welch, 1961; Rich and Black, 1964) and gives a more definitive value for exchangeable K (Merwin and Peech, 1950). However, $NH_4$ can fail to replace some of the K ions that are physically accessible to exchange until the $NH_4$ salt is added. In experiments with expanded Na-degraded mica particles that contained some interlayer K (but not enough to contract the mineral), it has been observed that the K can be replaced very easily by Na but not by $NH_4$ because the mineral contracts when it is placed in the $NH_4$ extracting solution. The degree of this effect in soils has not been determined.

The amounts of soil K that can be extracted with solutions of $NH_4$ salts and $NaBPh_4$ are obviously quite different. They both provide a measure of the exchangeability of K, but for different sets of conditions. A variety of other exchangeable K values that are usually somewhere between the $NH_4$- and $NaBPh_4$ levels can be determined with other replacing cations and procedures. Therefore, it is imperative that the determination and interpretation of exchangeable K values (and those for the other alkali elements) be based on clearly described methods.

In the 1979 *Glossary of Soil Science Terms* (Soil Science Society of America, Terminology Committees, 1979), exchangeable K has been characterized as "easily" exchanged, and often the rate with which soil K is extracted by neutral salts is one of the criteria used to distinguish between exchangeable and nonexchangeable forms. However, it should be emphasized that cation exchange is a diffusion process (Helfferich, 1962) and is subject to the rate-limiting effects of particle and film diffusion (Helfferich, 1966). Thus, many factors can be responsible for the rates observed for the exchange of soil K. Often, it is the method employed, and especially the restrictions imposed by the blocking effects of fixable cations, that slow the exchange of K in soil. Because an exchange of 1.6 to 6.5 cmol $kg^{-1}$ has been observed in 30 min with a group of Iowa soils (Smith *et al.*, 1968), it is apparent that much more than the $NH_4$-exchangeable K (<1 cmol $kg^{-1}$ in these soils) is subject to rapid exchange.

The literature contains numerous exchangeable K values that have been obtained with a wide variety of soils and several extraction procedures. Most of these values apply to soils that have been exposed to conventional tillage methods that mix fertilizer-applied K and crop residues with the surface soil. As these conventional methods give way to various conservation tillage systems (Unger and McCalla, 1980), profound changes in the distribution of exchangeable K in field soils can be expected. In the past, however, the *Soil Survey Investigations Reports* for different states (e.g., Soil Conservation Service–US Department of Agriculture, 1978) have been a particularly good source of information on the $NH_4$-exchangeable K levels in the different horizons of many United States soils. In general, the values reported for $NH_4$-exchangeable K vary from 0.1 to 1 cmol $kg^{-1}$ and comprise from 1 to 20% of the $NH_4$-exchangeable base content of the soil. On the other hand, relatively few determinations of the total exchangeable K in soils have been made. By using Na as a replacing cation and $NaBPh_4$ to precipitate the replaced K, however, 7.8 to 25.6 cmol K $kg^{-1}$ (20 to 47% of total K) has been shown to be exchangeable in 16 Iowa soils (Smith *et al.*, 1968). Differences in the rate of K exchange have been observed with these soils and their size fractions.

Air-dried soil samples are commonly used for exchangeable K determinations because they are convenient to handle. This practice continues even though it is well known that drying can cause appreciable changes in the $NH_4$-exchangeable K in many soils (Scott and Hanway, 1960). Both increases and decreases in $NH_4$-exchangeable K have been observed with drying and many factors have been shown to influence the degree of change. The curves in Figure 4 show the amount of K release that can occur with drying, the moisture levels at which the release occurs, and the reversion of released K (decrease in $NH_4$-exchangeable K level) that can occur when dried soils are rewet. Oven drying this Marshall subsoil at 110°C increased the $NH_4$-exchangeable K tenfold. Reversion in the rewet soil does not usually reduce the $NH_4$-exchangeable K level to the level in the field-moist soil. This effect of drying is mainly a problem when soils are dried for laboratory analyses or greenhouse experiments. However, an effect of drying has been observed in the top 2.5 cm of soils in the field (Luebs *et al.*, 1956). In any event, many of the $NH_4$-exchangeable

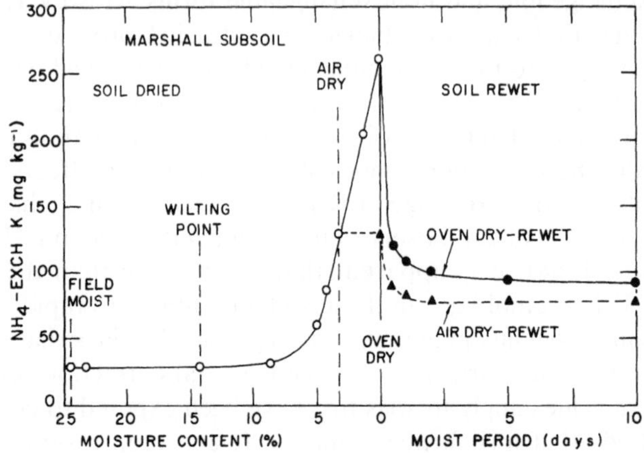


**Figure 4.** Ammonium-exchangeable K in Marshall subsoil samples that were dried to different moisture levels or were dried and rewet in $H_2O$ for different periods.

values for soil K that have been determined with dry soil samples should be used with caution. Even though there are ways of controlling the release and reversion of K associated with changes in soil moisture (Bates and Scott, 1969), the $NH_4$-exchangeable K existing in field-moist soil is best determined with undried soil samples. Drying has relatively little effect on the amount of exchangeable K determined with NaCl–$NaBPh_4$ solutions (Smith *et al.*, 1968).

In a review of the literature on exchangeable Na in humid-region soils, Larson (1949) explained the lack of attention to the subject on the basis of the very small amounts of exchangeable Na in soils with good drainage, the lack of evidence of any great importance of the small amounts present, and the difficulty with which the small amounts were determined. The analytical difficulties no longer exist but the lack of emphasis on exchangeable Na in these soils continues. Nevertheless, many more values for this form of Na are now available in the *Soil Survey Investigations Reports* for different states (e.g., Soil Conservation Service–US Department of Agriculture, 1978). In the A horizon of 42 soils studied by Larson and Allaway (1950) the $NH_4$-exchangeable Na varied from 0.03 to 0.27 cmol $kg^{-1}$ and averaged 0.13 cmol $kg^{-1}$. Similar levels have been reported for all depths in the profile of many soils but much higher concentrations can exist if drainage is restricted. For instance, $NH_4$-exchangeable Na values of up to 2 cmol $kg^{-1}$ have been obtained with samples from the $B_3$ horizon of Edina soil (Larson, 1949). In arid regions, the surface soils can have $NH_4$-exchangeable Na levels comparable to those of humid-region soils, but the subsoils often have higher concentrations (e.g., 2.8 cmol Na $kg^{-1}$ in the $C_2$ horizon of Wiley soil in Colorado). Of course, salt-affected soils can have very high $NH_4$-exchangeable Na levels to the surface. An exchangeable sodium percentage (ESP) value of 15% has been used as the level at which the deleterious effects of Na begin but the current classification of soil salinity has shifted toward the use of the SAR (Bresler *et al.*, 1982).

Soil Na that occurs with K in the interlayers of contracted layer silicates is subject to exchange under the conditions required for the exchange of K. The results of a comparison of the expanding tendencies of ground paragonite and muscovite (Barshad, 1950) suggest the interlayer Na is more susceptible to exchange than K. On the other hand, interlayer Na in fully contracted samples of K-depleted micas is very difficult to displace because the basal spacing is less than that of the original mica. The exchange of Na in soil zeolites may be restricted by ion sieving (Sherry, 1969). In soils that contain analcime, therefore, determinations of exchangeable Na can be complicated by differences in the results obtained with various replacing cations (Baldar and Whittig, 1968; Babcock and Schultz, 1970).

Relatively few determinations of the exchangeable fraction of indigenous Li in soils have been carried out. Using neutral normal $NH_4OAc$, Bradford and Pratt (1961) found the extractable Li in 12 soils from different areas of California to range from 0.1 to 0.9 mg $kg^{-1}$ and average 0.3 mg $kg^{-1}$. Likewise, ranges of 0.002 to 0.409 and 0.032 to 0.830 mg $kg^{-1}$ have been reported by Davey and Wheeler (1980) for levels of $NH_4$-exchangeable Li in 50 Papua, New Guinea, and 5 Australian soils, respectively. On the other hand, there has been a great deal of interest in the behavior of Li as an exchangeable cation in smectites because there are various reports that Li can migrate into empty octahedral sites of the clay (Prost and Calvet, 1969). Whereas much of the experimental data has led to the conclusion that Li does migrate to these octahedral sites when the smectite is heated at 200 to 300°C, Tettenhorst (1962) concluded the Li cation is only in the hexagonal holes of the silicate surface.

### C. Nonexchangeable

The literature on the nonexchangeable K in soils is very extensive (H.W. Martin and Sparks, 1985), but it is clear that this fraction of the soil K is ill defined. Talibudeen (1972) has said the term "nonexchangeable" K is a misnomer and Metson (1980) has pointed out that concepts of "difficultly exchangeable" and "slowly exchangeable" are often involved. The problem, of course, centers on the previously discussed question of what constitutes a clearly defined exchangeable fraction of the soil K. In addition, it should be noted that different degrees of exchangeability are assigned to K that is associated with specific sites in the soil (Metson, 1980). Moreover, the wedge sites at the frayed edges of K-depleted mica particles (Fanning and Keramidas, 1977) are deemed to be particularly responsible for K being nonexchangeable (H.W. Martin and Sparks, 1985). The curves in Figure 2 do not support these views. Instead, the selectivity of the mica for K increased in a continuous manner as the degree of K depletion increased. Therefore, it is difficult to see how a particular portion of this interlayer K can be segregated as a nonexchangeable form.

Following the concepts developed with the exchangeable alkalies, only those alkali cations in soils that are not potentially exchangeable with other cations should be categorized as nonexchangeable. However, much more of the alkalies is usually reported as nonexchangeable because the methods adopted for the exchangeable cation determinations fail to include all the cations they should. Even a portion of the alkali cations located on exposed surfaces of soil solids may appear to be nonexchangeable if a single equilibrium extraction is made with an inadequate concentration of replacing cation. Less obvious discrepancies in nonexchangeable alkali determinations arise when the method of extracting the exchangeable cations causes a contraction of otherwise expanded layer silicates, fails to promote the expansion of contracted layer silicates, or utilizes replacing cations that are too large to enter soil zeolite structures.

Some of the alkali cations in soils are nonexchangeable by virtue of their location in the solid fraction. Most of the alkali cations in feldspar particles are included in this category. Various observations of cation-exchange reactions on the surfaces of the feldspar particles have been made, but prolonged treatments of orthoclase with NaCl–$NaBPh_4$ solutions have shown that an exchange of cations within the particles does not occur (Smith *et al.,* 1968). The exchange of Li in the octahedral layers of micaceous minerals and of alkali cations in pyroxenes or amphiboles is probably confined to outer particle surfaces as well. Some of the interlayer cations in small mica particles may be nonexchangeable (Scott, 1968) but this situation is likely to change when the reasons for the discrepency between coarse and fine particles are better known. As long as plant and animal tissues in the soil remain alive most of their content of alkali elements can be considered nonexchangeable. Organic matter and amorphous inorganic coatings on soil particles may be responsible for some of the alkali cations being inaccessible to exchange, but limited experiments with illitic materials have not substantiated this possibility (Smith, 1967).

### D. Fixed

It is often convenient to delineate a portion of the alkali cations in soils that is called "fixed" but it is evident from the extensive reviews of the literature on the fixation of K and other cations (Reitemeier, 1951; Schuffelen and van der Marel, 1955; Agarwal, 1960; Sawhney, 1972) that many details about this form of the alkalies are still lacking. In general, the concept of cation fixation is based on the idea that some cations in soils are sorbed more strongly than others and may resist replacement by other cations to such a degree that they are said to be fixed (Sawhney, 1972).

Additional amounts of the alkalies can be fixed by being incorporated in newly formed minerals (e.g., K in jarosite). In either event, the term "fixation" has been used to indicate a decrease in the extractability of cations that were otherwise in a soluble or exchangeable state in the soil. Often, cation fixation experiments focus on the fate of added cations but similar changes can occur in the indigenous cations in the soil (e.g., a decrease in exchangeable K, instead of the increase shown in Figure 4, when soils are dried). To make these determinations of cation fixation, some method of extraction must be adopted. Many methods have been used and the results have varied accordingly.

Most of the research on the fixation of alkali cations has been carried out with K and has dealt with the fixation of this cation between the layers of phyllosilicates, but the same principles apply to Rb and Cs. Explanations for this fixation have been given by Eberl (1980) and Sawhney (1972). With any interlayer cation, fixation depends in part on the fact that the mineral contracts when the force of attraction between the cation and the mineral exceeds that of the cation for its hydration shell. This is what happens when there is enough K, Rb, or Cs in the interlayers of minerals with a relatively high layer charge. However, the interlayer cations in contracted layer silicates are not fixed unless the potential replacing cations in the surrounding solution phase are unable to expand the mineral. Indeed, the same concepts as have been presented for exchangeable alkali forms are applicable in this situation in that the interlayer cations may be fixed against $NH_4$ exchange but still be completely exchangeable with Na. For this reason, the fixation of alkali cations in contracted layer silicates must be defined in terms of the replacing cations or procedures that do not expand the mineral. Unfortunately, interlayer alkali cations in expanded layer silicates can appear to be fixed as well if the replacing cation is $NH_4$ because high concentrations of the added $NH_4$ can contract the layers. Other processes can also cause changes in the exchangeability of alkali cations in soils (Black, 1968), but these changes should be unaffected by the nature of the replacing cation unless ion sieving in zeolites is involved.

Indigenous and fixed (added) cations in the interlayers of contracted layer silicates appear to be comparable forms in that they are not exchangeable with $NH_4$ but can be replaced completely by Na or other cations that tend to expand the mineral. Differences in their rates of exchange with Na, however, show they are not. This distinction is shown in Figure 5 with some K release results that were obtained with samples of biotite and K-degraded biotite. In this experiment, the conditions for K exchange were optimized for both samples by using NaCl–$NaBPh_4$ extracting solutions, but the added K was replaced much faster than the indigenous K in biotite. This difference did not change when the samples were oven dried to insure the removal of interlayer $H_2O$. The degree of interlayer cation exchange by hydrated cations is obviously an inadequate criterion for the delineation of alkali forms.

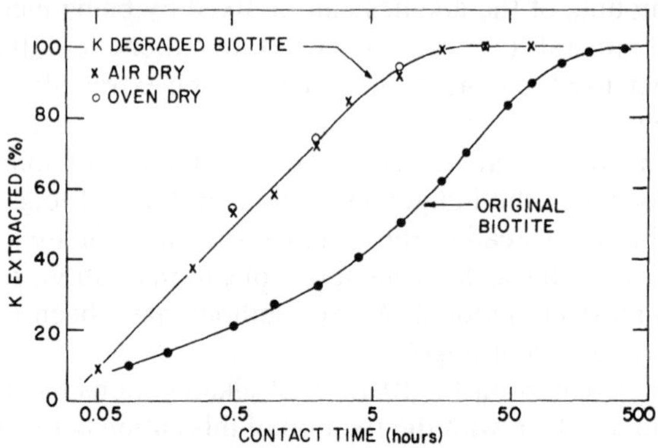


**Figure 5.** Potassium extracted from 10- to 20- μm samples of biotite and K-degraded biotite (K-depleted mica that was resaturated with K and air dried or oven dried at 110°C) by different contact periods with NaCl–$NaBPh_4$ solutions at 25°C.

## References


Addiscott, T.M., and O. Talibudeen. 1969. The buffering capacity of potassium reserves in soils. *Potash Rev.* 4/45. 24pp.

Agarwal, R.R. 1960. Potassium fixation in soils. *Soils Fertil.* 23:375–378.

Ahrens, L.H. 1964. The significance of the chemical bond for controlling the geochemical distribution of the elements. Part 1. *Phys. Chem. Earth* 5:1–54.

Ahrens, L.H. 1965. *Distribution of the Elements in our Planet.* McGraw-Hill Book Co., New York.

Aidinyan, R. Kh. 1959. Distribution of rare alkalies in soil colloids and the participation of plants in this process. *Geochemistry* 4:428–441.

Aller, L.H. 1961. *The Abundance of the Elements.* Wiley-Interscience, New York.

American Geological Institute. 1976. *Glossary of Geology and Related Sciences.* Anchor Press, New York.

Angino, E.E., H.L. Cannon, K.M. Hambidge, and A.W. Voors. 1974. Lithium. In: W. Mertz (ed.) *Geochemistry and the Environment.* Vol. 1. National Academy of Sciences, Washington, DC, pp. 36–42.

Arnold, M. A., and M. E. Meyerhoff. 1984. Ion-selective electrodes. *Anal. Chem.* 56:20R–48R.

Asher, C. J., and P. G. Ozanne. 1967. Growth and potassium content of plants in solution cultures maintained at constant potassium concentrations. *Soil Sci.* 103:155–161.

Aubert, H., and M. Pinta. 1977. *Trace Elements in Soils.* Elsevier, New York.

Babcock, K. L., and R. L. Schultz. 1970. Isotopic and conventional determination of exchangeable sodium percentage of soil in relation to plant growth. *Soil Sci.* 109:19–22.

Baldar, W. A., and L. D. Whittig. 1968. Occurrence and synthesis of soil zeolites. *Soil Sci. Soc. Am. Proc.* 32:235–238.

Barber, S. A. 1984. *Soil Nutrient Bioavailability.* John Wiley and Sons, New York.

Barber, S. A., J. M. Walker, and E. H. Vasey. 1963. Mechanisms for the movement of plant nutrients from the soil and fertilizer to the plant root. *J. Agric. Food Chem.* 11:204–207.

Barshad, I. 1950. The effect of the interlayer cations on the expansion of the mica type of crystal lattice. *Am. Mineral.* 35:225–228.

Barshad, I. 1964. Chemistry of soil development. In: F. E. Bear (ed.) *Chemistry of the Soil.* 2nd ed. Reinhold, New York, pp. 1–70.

Barth, T. W. 1969. *Feldspars.* Wiley-Interscience, New York.

Bassett, W. A. 1959. The origin of the vermiculite deposit at Libby, Montana. *Am. Mineral.* 44:282–299.

Bates, T. E., and A. D. Scott. 1969. Control of potassium release and reversion associated with changes in soil moisture. *Soil Sci. Soc. Am. Proc.* 33:566–568.

Baver, L. D., and N. S. Hall. 1937. Colloidal properties of soil organic matter. Res. Bull. 267. Missouri Agricultural Experiment Station.

Bear, F. E. (ed.) 1953. Sodium symposium. *Soil Sci.* 76:1–96.

Beckett, P. H. T. 1971. Potassium potentials—a review. *Potash Rev.* 5/30, 41 pp.

Bingham, F. T., A. L. Page, and C. R. Bradford. 1964. Tolerance of plants to lithium. *Soil Sci.* 98:4–8.

Black, C. A. 1968. *Soil-Plant Relationships.* 2nd ed. John Wiley and Sons, New York.

Bolton, J. 1966. Distribution and rate of release of cations from mechanical fractions of soils. *Ann. Rept. Rothamsted Expt. Sta. 1965*, pp. 61–62.

Bolton, J. 1971. Quantity-intensity relationships for labile sodium in field soils. *J. Soil Sci.* 22:417–429.

Bouat, M. 1969. L'utilisation du potassium 40 et potassium 42 en agronomie. *Ann. Agron.* 20:89–104.

Bowen, H. J. M. 1979. *Environmental Chemistry of the Elements.* Academic Press, New York.

Bowen, H. J. M., and P. A. Cawse. 1965. Some effects of gamma radiation on the composition of the soil solution and soil organic matter. *Soil Sci.* 98:358–361.

Boyer, J. 1972. Soil potassium. In: *Soils of the Humid Tropics.* National Academy of Sciences, Washington, DC, pp. 102–135.

Bradford, G. R. 1963. Lithium survey of California's water resources. *Soil Sci.* 96:77–81.

Bradford, G. R. 1966. Lithium. In: H. D. Chapman (ed.) *Diagnostic Criteria for Plants and Soils.* University of California, Division of Agricultural Sciences, pp. 218–224.

Bradford, G. R., and P. F. Pratt. 1961. Separation and determination of lithium in irrigation water, plant material, and soil extracts. *Soil Sci.* 91:189–193.

Brady, N. C. 1984. *The Nature and Properties of Soils.* 9th ed. Macmillan, New York.

Bresler, E., B. L. McNeal, and D. L. Carter. 1982. *Saline and Sodic Soils.* Springer-Verlag, Berlin.

Briscoe, H. V. A., A. A. Eldridge, G. M. Dyson, and A. J. E. Welch (eds.) 1961. *Mellor's Comprehensive Treatise on Inorganic and Theoretical Chemistry.* Vol. II, Suppl. II. John Wiley and Sons, New York.

Briscoe, H. V. A., A. A. Eldridge, G. M. Dyson, and A. J. E. Welch (eds.) 1963. *Mellor's Comprehensive Treatise on Inorganic and Theoretical Chemistry.* Vol. II, Suppl. III. John Wiley and Sons, New York.

Brownell, P. F. 1979. Sodium as an essential micronutrient element in plants and its role in metabolism. *Adv. Bot. Res.* 7:117–224.

Buckman, H. O., and N. C. Brady. 1969. *The Nature and Properties of Soils.* 7th ed. Macmillan, New York.

Burbidge, E. M., G. R. Burbidge, W. A. Fowler, and F. Hoyle. 1957. Synthesis of the elements in stars. *Rev. Mod. Phys.* 29:547–650.
Burridge, J. C., and P. M. Ahn. 1965. A spectrographic survey of representative Ghana forest soils. *J. Soil Sci.* 16:296–309.
Butler, J. R. 1954. Trace-element distribution in some Lancashire soils. *J. Soil Sci.* 5:156–166.
Call, F. 1961. Biological properties of lithium. In: H.V.A. Briscoe, A.A. Eldridge, G.M. Dyson, and A.J.E. Welch (eds.) *Mellor's Comprehensive Treatise on Inorganic and Theoretical Chemistry.* Vol. II, Suppl. II. John Wiley and Sons, New York, pp. 293–304.
Chester, R. 1965. Elemental geochemistry of marine sediments. In: J.P. Riley and G. Skirrow (eds.) *Chemical Oceanography.* Vol. 2. Academic Press, New York, pp. 23–80.
Clarke, F. W. 1924. The data of geochemistry. US Geological Survey Bull. 770.
Coffey, G. N. 1912. A study of the soils of the United States. US Department of Agriculture, Bureau of Soils, Bull. No. 85.
Cook, M. G., and C. I. Rich. 1962. Weathering of sodium-potassium mica in soils of the Virginia Piedmont. *Soil Sci. Soc. Am. Proc.* 26:591–595.
Cooke, G. W. 1967. *The control of Soil Fertility.* Lockwood, London.
Cotton, F. A., and G. Wilkinson. 1980. *Advanced Inorganic Chemistry: A Comprehensive Text.* John Wiley and Sons, New York.
Coughtrey, P. J., and M. C. Thorne. 1983a. Rubidium. In: *Radionuclide Distribution and Transport in Terrestial and Aquatic Ecosystems. A Critical Review of Data.* Vol. 1. A.A. Balkema, Rotterdam, pp. 69–92.
Coughtrey, P. J., and M. C. Thorne. 1983b. Cesium. In: *Radionuclide Distribution and Transport in Terrestial and Aquatic Ecosystems. A Critical Review of Data.* Vol. 1. A.A. Balkema, Rotterdam, pp. 321–424.
Coughtrey, P. J., D. Jackson, and M. Thorne. 1983. Sodium. In: *Radionuclide Distribution and Transport in Terrestial and Aquatic Ecosystems. A Critical Review of Data.* Vol. 3. A.A. Balkema, Rotterdam, pp. 1–41.
Curtis, C. D., P. E. Brown, and V. A. Somogyi. 1969. A naturally occurring sodium vermiculite from Unst, Shetland. *Clay Mineral.* 8:15–19.
Davey, B. G., and R. C. Wheeler. 1980. Some aspects of the chemistry of lithium in soils. *Plant Soil* 57:49–60.
Davis, J. J. 1963. Cesium and its relationship to potassium in ecology. In: V. Schultz and A.W. Klement (eds.) *Radioecology.* Reinhold, New York.
Day, F. H. 1964. *The Chemical Elements in Nature.* Reinhold, New York.
Deer, W. A., R. A. Howie, and J. Zussman. 1962. *Rock Forming Minerals,* Vol. 3. *Sheet Silicates.* Longmans, London.
Diamond, J. M., and E. M. Wright. 1969. Biological membranes: the physical basis of ion and nonelectrolyte selectivity. *Ann. Rev. Physiol.* 31:581–646.
Diest, J., and O. Talibudeen. 1967. Rubidium-86 as a tracer for exchangeable potassium in soils. *Soil Sci.* 104:119–122.
Donnay, G., and J. W. Gryder. 1964. The ionic radius of lithium. *Carnegie Inst. Wash. Year Book* 63:238–239.
Duthion, C. 1968. Potassium in the soil. *Potash Rev.* 4/43. 21pp.
Eaton, F. M., and V. P. Sokoloff. 1935. Absorbed sodium in soils as affected by the soil-water ratio. *Soil Sci.* 40:237–247.
Eberl, D. D. 1980. Alkali cation selectivity and fixation by clay minerals. *Clays Clay Mineral.* 28:161–172.

Eckert, D. J., and E. O. McLean. 1980. Differential bonding of potassium and rudibium-86 in soils of differing clay type and degree of weathering. *Soil Sci. Soc. Am. J.* 44:425–428.

Eisenman, G. 1962. Cation-selective glass electrodes and their mode of operation. *Biophys. J.* 2(part 2):259–324.

Eisenman, G. 1969. Theory of membrane electrode potentials: an examination of the parameters determining the selectivity of solid and liquid ion exchangers and of neutral sequestering molecules. In: R.A. Durst (ed.) *Ion-Selective Electrodes*. National Bureau of Standards, Spec. Publ. 314, pp. 1–56.

Epstein, E. 1960. Calcium-lithium competition in absorption by plant roots. *Nature* (London) 1985:705–706.

Evans, D. W., J. J. Alberts, and R. A. Clark III. 1983. Reversible ion-exchange fixation of cesium-137 leading to mobilization from reservoir sediments. *Geochim. Cosmochim. Acta* 47:1041–1049.

Failyer, G. H., J. G. Smith, and H. R. Wade. 1908. The mineral composition of soil particles. US Department of Agriculture, Bureau of Soils, Bull. No. 54.

Fanning, D. S., and V. Z. Keramidas. 1977. Micas. In: J.B. Dixon and S. B. Weed (eds.) Minerals in soil environments. Soil Science Society of America, Madison, WI, pp. 195–258.

Farrell, R. E. 1979. Electrochemical determination of extractable potassium in micaceous minerals. M.S. Thesis. Iowa State University, Ames, IA.

Farrell, R. E. 1985. Development and application of potentiometric methods of characterizing potassium in soils and micaceous minerals. Ph.D. Diss. Iowa State University, Ames, IA. (Diss. Abstr. 85-24650).

Foster, N. D. 1960. Interpretation of the composition of lithium micas. US Geological Survey, Prof. Paper 354-E.

Frere, M. H., R. G. Menzel, K. H. Larson, R. Overstreet, and R. F. Reitemeier. 1963. The behavior of radioactive fallout in soils and plants. National Academy of Sciences–National Research Council, Publ. 1092.

Gentili, R. 1954. The geochemistry of potassium. *Potassium Symp.* 1954:27–40.

Goldschmidt, V. M. 1954. *Geochemistry*. Clarendon Press, Oxford.

Goles, G. G. 1969. Cosmic abundances. In: K.H. Wedepohl (ed.) *Handbook of Geochemistry*. Vol. I. Springer-Verlag, Berlin, pp. 116–133.

Gordy, W., and W. J. O. Thomas. 1956. Electronegativities of the elements. *J. Chem. Phys.* 24:439–444.

Goulding, K. W. T. 1983. Thermodynamics and potassium exchange in soils and clay minerals. *Adv. Agron.* 36:215–264.

Greenwood, R. 1960. Availability of cesium for ion rockets. *Mining Eng.* 12:482–483.

Grimes, D. W. 1966. An evaluation of the availability of potassium in crop residues. Ph.D. Diss. Iowa State University, Ames, IA (Diss. Abstr. 66-06981).

Grimme, H., and K. Nemeth. 1978. The evaluation of soil K status by means of soil testing. *Proc. 11th Congr. Int. Potash Inst.* 1978:99–108.

Hanway, J. J. 1954. Fixation and release of ammonium in soils and certain minerals. Ph.D. Diss. Iowa State University, Ames, IA.

Hanway, J. J., and A. D. Scott. 1956. Ammonium fixation and release in certain Iowa soils. *Soil Sci.* 82:379–386.

Hanway, J. J., A. D. Scott, and G. Stanford. 1957. Replaceability of ammonium fixed in clay minerals as influenced by ammonium or potassium in the extracting solution. *Soil Sci. Soc. Am. Proc.* 21:29–34.

Hargett, N. L., and J. T. Berry. 1985. Fertilizer summary data. TVA Bull. 189. Muscle Shoals, AL.
Heier, K. S., and J. A. S. Adams. 1964. The geochemistry of the alkali metals. *Phys. Chem. Earth* 5:253–381.
Heier, K. S., and G. K. Billings. 1970. In: K.H. Wedepohl (ed.) *Handbook of Geochemistry*. Vols. II/1, II/2, and II/3. Springer-Verlag, Berlin.
Helfferich, F. 1962. *Ion Exchange*. McGraw-Hill Book Co., New York.
Helfferich, F. 1966. Ion-exchange kinetics. In: J.A. Marinsky (ed.) *Ion Exchange*. Vol. 1. Marcel Dekker, New York, pp. 65–100.
Heydemann, A. 1969. Tables. In: K.H. Wedepohl (ed.) *Handbook of Geochemistry*. Vol. 1. Springer-Verlag, Berlin, pp. 376–412.
Hood, J. T., N. C. Brady, and D. J. Lathwell. 1956. The relationship of water soluble and exchangeable potassium to yield and potassium uptake by Ladino clover. *Soil Sci. Soc. Am. Proc.* 20:228–231.
Horstman, E. L. 1957. The distribution of lithium, rubidium, and caesium in igneous and sedimentary rocks. *Geochim. Cosmochim. Acta* 12:1–28.
International Potash Institute. 1972. Mineralogy of soil potassium. In: *Potassium in Soil. Proc. 9th Colloq. Int. Potash Inst.* Landshut, Federal Republic of Germany, pp. 13–71.
Jackson, M. L. 1964. Chemical composition of soils. In: F.E. Bear (ed.) *Chemistry of the Soil*. 2nd ed. Reinhold, New York, pp. 71–141.
Jeffries, C. D. 1947. The mineralogical approach to some soil problems. *Soil Sci.* 63:315–320.
Johnson, F. N. 1984. *The Psychopharmacology of Lithium*. MacMillan Press, London.
Kabata-Pendias, A., and H. Pendias. 1984. *Trace Elements in Soils and Plants*. CRC Press, Boca Raton, FL.
Kelley, W. P. 1951. *Alkali Soils, Their Formation, Properties and Reclamation*. Reinhold, New York.
Kilmer, V. J. 1965. Silicon. *Agronomy* 9:959–962.
Kirkham, M. B. 1979. Sludge disposal. In: R.W. Fairbridge and C.W. Finkl, Jr. (eds.) *Encyclopedia of Soil Science*. Part I. Dowden, Hutchinson & Ross, Stroudsburg, PA, pp. 429–433.
Lange, I. M., R. C. Reynolds, and J. B. Lyons. 1966. K/Rb ratios in coexisting K-feldspars and biotites from some new England granites and metasediments. *Chem. Geol.* 1:317–322.
Larson, W. E. 1949. Release of sodium from nonreplaceable to replaceable forms in Iowa soils and the response of various crops to sodium fertilization. Ph.D. Diss. Iowa State University, Ames, IA.
Larson, W. E., and W. H. Allaway. 1950. Release of sodium from nonreplaceable to replaceable forms in some Iowa soils. *Soil Sci.* 70:249–256.
Lee, S. Y., and R. W. Tank. 1985. Role of clays in the disposal of nuclear waste: A review. *Appl. Clay Sci.* 1:145–162.
Liu, M. M. 1970. Exchangeability of potassium in Marshall soil. M.S. Thesis. Iowa State University, Ames, IA.
Livingstone, D. A. 1963. Chemical composition of rivers and lakes. US Geological Survey, Prof. Paper 440-G.
Lomenick, T. F., D. G. Jacobs, and E. G. Struxness. 1967. The behavior r of strontium-90 and cesium-137 in seepage pits at ORNL. *Health Phys.* 13:897–905.

Loughnan, F. C. 1969. *Chemical Weathering of the Silicate Minerals.* Elsevier, New York.

Luebs, R. E., G. Stanford, and A. D. Scott. 1956. Relation of available potassium to soil moisture. *Soil Sci. Soc. Am. Proc.* 20:45–50.

Maddock, A. G. 1963. Francium. In: H.V. Briscoe, A.A. Eldridge, G.M. Dyson, and A.J.E. Welch (eds.) *Mellor's Comprehensive Treatise on Inorganic and Theoretical Chemistry.* Vol. II, Suppl. III. John Wiley and Sons, New York, pp. 2506–2521.

Marbut, C. F. 1935. Soils of the United States. In: *Atlas of American Agriculture.* Part 3. US Department of Agriculture, Washington, DC.

Martin, G. R. 1963. The natural radioactivity of the alkali metals. In: H.V.A. Briscoe, A.A. Eldridge, G.M. Dyson and A.J.E. Welch (eds.) *Mellor's Comprehensive Treatise on Inorganic and Theoretical Chemistry.* Vol. II, Suppl. III. John Wiley and Sons, New York, pp. 2408–2505.

Martin, H. W., and D. L. Sparks. 1985. On the behavior of nonexchangeable potassium in soils. *Commun. Soil Sci. Plant Anal.* 16:133–162.

Mason, B. 1966. *Principles of Geochemistry.* 3rd ed. John Wiley and Sons, New York.

Matthews, B.C., and J.A. Smith. 1957. A percolation method for measuring potassium-supplying power of soils. *Can. J. Soil Sci.* 37:21–28.

McCallan, M.E., B.M. O'Leary, and C.W. Rose. 1980. Redistribution of cesium-137 by erosion and deposition on an Australian soil. *Austr. J. Soil Res.* 18:119–128.

McGeorge, W.T. 1931. Organic compounds associated with base exchange reactions in soils. Arizona Agr. Expt. Sta. Tech. Bull. No. 31.

McHenry, J.R., and G.D. Bubenzer. 1985. Field erosion estimated from $^{137}Cs$ activity measurements. *Trans. Am. Soc. Agric. Eng.* 28:480–483.

McLaughlin, R.J.W. 1955. Geochemical changes due to weathering under varying climatic conditions. *Geochim. Cosmochim. Acta* 8:109–130.

McLaughlin, R.J.W. 1958. Geochemical partition in two illitic clays. *Geochim. Cosmochim. Acta* 15:165–169.

McLaughlin, R.J.W. 1959. The geochemistry of some kaolinitic clays. *Geochim. Cosmochim. Acta* 17:11–16.

Mellor, J.W. 1922. *A Comprehensive Treatise on Inorganic and Theoretical Chemistry.* Vol. II. Longmans Green, London.

Mengel, K., and E.A. Kirby. 1980. Potassium in crop production. *Adv. Agron.* 33:59–103.

Menzel, R.G., and S.J. Smith. 1984. Soil fertility and plant nutrition. In: M.F. L'Annunziata and J.O. Legg (eds.) *Isotopes and Radiation in Agricultural Sciences.* Vol. 1. Academic Press, London, pp. 1–34.

Merwin, H.D., and M. Peech. 1950. Exchangeability of soil potassium in the sand, silt, and clay fractions as influenced by the nature of the complementary exchangeable cation. *Soil Sci. Soc. Am. Proc.* 15:125–128.

Metson, A.J. 1980. Potassium in New Zealand soils. New Zealand Soil Bureau, Sci. Rept. 38. 61 pp.

Mielniczuk, J. 1979. Forms of potassium in Brazilian soils. *Potash Rev.* 4/63. 13 pp.

Mitchell, R.L. 1964. Trace elements in soils. In: F.E. Bear (ed.) *Chemistry of the Soil.* 2nd ed. Reinhold, New York, pp. 320–368.

Mortensen, J.L., and F.L. Himes. 1964. Soil organic matter. In: F.E. Bear (ed.) *Chemistry of the Soil.* 2nd ed. Reinhold, New York, pp. 206–241.

Moss, P. 1969. A comparison of potassium-activity ratios derived from equilibration procedures and from measurements on displaced soil solution. *J. Soil Sci.* 20:297–306.

Norrish, K. 1972. Factors in the weathering of mica to vermiculite. *Proc. 1972 Int. Clay Conf.* 2:83–101.

Norton, J.J. 1965. Lithium-bearing bentonite deposit, Yavapai County, Arizona. pp. D163–D166. In: Geological survey research 1965. US Geological Survey, Prof. Paper 525-D.

Nye, P.H. 1972. Localized movement of potassium ions in soil. *Proc. 9th Colloq. Int. Potash Inst.* pp. 147–155.

Parker, F.W. 1921. Methods of studying the concentration and composition of the soil solution. *Soil Sci.* 12:209–232.

Parker, R.L. 1967. Data of geochemistry. Chapter D. Composition of the earth's crust. US Geological Survey, Prof. Paper 440-D.

Pearson, G.A. 1960. Tolerance of crops to exchangeable sodium. US Department of Agriculture, Information Bull. 216.

Pettijohn, F.J. 1957. *Sedimentary Rocks*. 2nd ed. Harper & Row, New York.

Prost, R., and R. Calvet. 1969. Position du lithium dans une montmorillonite Li chauffee. *Compt. Rend. Acad. Sci.* 269D:539–541.

Quemener, J. 1979. The measurement of soil potassium. IPI Research Topics. No. 4. Intern. Potash Institute. Bern-Worblanfen, Switzerland.

Rankama, K., and Th.G. Sahama. 1950. *Geochemistry*. University of Chicago Press, Chicago.

Rausell-Colom, J.A., T.R. Sweatman, C.B. Wells, and K. Norrish. 1965. Studies in the artificial weathering of mica. In: E.G. Hallsworth and D.V. Crawford (eds.) *Experimental Pedology*. Butterworths, London. pp. 40–72.

Reichenberg, D. 1966. Ion-exchange selectivity. In: J.A. Marinsky (ed.) *Ion exchange*. Vol. I. Marcel Dekker, New York. pp. 227–276.

Reitemeier, R.F. 1946. Effect of moisture content on the dissolved and exchangeable ions of soils of arid regions. *Soil Sci.* 61:195–214.

Reitemeier, R.F. 1951. Soil potassium. *Adv. Agron.* 3:113–164.

Reitemeier, R.F. 1957. Soil potassium and fertility. In: A. Stefferud (ed.), *Soil. 1957 Yearbook of Agriculture.* US Government Printing Office, Washington, DC, pp. 101–106.

Rich, C.I. 1968. Mineralogy of soil potassium. In: V.J. Kilmer, S.E. Younts, and N.C. Brady (eds.) *The Role of Potassium in Agriculture*. American Society of Agronomy, Crop Science Society of America, Soil Science Society of Agronomy, Madison, WI, pp. 79–108.

Rich, C.I., and W.R. Black. 1964. Potassium exchange as affected by cation size, pH, and mineral structure. *Soil Sci.* 97:384–390.

Richards, F.A. 1956. On the state of our knowledge of trace elements in the ocean. *Geochim. Cosmochim. Acta* 10:241–243.

Richter, D. 1965. Potassium in soil and plant. Part 1. Occurrence and behavior of potassium in soils. *Fortschrittskerichte fur die Landwirtschaft.* 14. 47 pp.

Ritchie, J.C., and J.R. McHenry. 1973. Vertical distribution of fallout cesium-137 in cultivated soils. *Radiation Data Repts.* 14:727–728.

Ritchie, J.C., J.R. McHenry, A.C. Gill and P.H. Hanks. 1970. The use of cesium-137 as a tracer of sediment movement and deposition. *Proc. Miss. Water Res. Conf.* pp. 149–162.

Ritchie, K.D. (ed.) 1979. K fertility in oxisols and ultisols of the humid tropics. Cornell International Agricultural Bull. 37. 45 pp.

Robert, J.L., M. Volfinger, J.N. Barrandon, and M. Basutcu. 1983. Lithium in the interlayer space of synthetic trioctahedral micas. *Chem. Geol.* 40:337–351.

Robinson, B.P. 1962. Ion-exchange minerals and disposal of radioactive wastes—A survey of literature. US Geological Survey Water-Supply Paper 1616.

Rosseinsky, D.R. 1965. Electrode potentials and hydration energies. Theories and correlations. *Chem. Rev.* 65:467–490.

Salmon, R.C. 1964. Potassium in different fractions of some Rhodesian soils. *Rhodesian J. Agric. Res.* 2:85–90.

Sawhney, B.L. 1972. Selective sorption and fixation of cations by clay minerals: A review. *Clays Clay Mineral.* 20:93–100.

Schroeder, D. 1978. Structure and weathering of potassium containing minerals. *Proc. 11th Congr. Int. Potash Inst.* 1978:5–25.

Schuffelen, A.C., and H.W. van der Marel. 1955. Potassium fixation in soils. *Potassium Symp.* 1955:157–201.

Schultz, R.K., R. Overstreet, and I. Barshad. 1960. On the soil chemistry of cesium 137. *Soil Sci.* 89:16–27.

Scott, A.D. 1968. Effect of particle size on interlayer potassium exchange in micas. *Trans. 9th Int. Congr. Soil Sci.* 2:649–660.

Scott, A.D., and J.J. Hanway. 1960. Factors influencing the change in exchangeable soil K observed on drying. *Trans. 7th Int. Congr. Soil Sci.* 3:72–79.

Scott, A.D., and S.J. Smith. 1966. Susceptibility of interlayer potassium in micas to exchange with sodium. *Clays Clay Mineral.* 14:69–81.

Scott, A.D., and L.F. Welch. 1961. Release of nonexchangeable soil potassium during short periods of cropping and sodium tetraphenylboron extraction. *Soil Sci. Soc. Am. Proc.* 25:128–132.

Shainberg, I., and W.D. Kemper. 1966. Electrostatic forces between clay and cations as calculated and inferred from electrical conductivity. *Clays Clay Mineral.* 14:117–132.

Shainberg, I. and W.D. Kemper, 1967. Ion exchange equilibria on montmorillonite. *Soil Sci.* 103:4–9.

Sharpley, A.N., S.J. Smith, R.G. Menzel, and R.L. Westerman. 1985. The chemical composition of rainfall in the Southern Plains and its impact on soil and water quality. Oklahoma State University Tech. Bull. T-162.

Sherry, H.S. 1969. The ion-exchange properties of zeolites. In: J.A. Marinsky (ed.) *Ion Exchange*. Vol. 2. Marcel Dekker, New York, pp. 89–133.

Short, N.M. 1961. Geochemical variations in four residual soils. *J. Geol.* 69:534–571.

Shukla, U.C., and K.G. Prasad. 1973. Forms and distribution of lithium, boron and fluorine in some sierozem soils of Haryama. *Indian J. Agric. Sci.* 43:934–937.

Smith, S.J. 1967. Susceptibility of interlayer potassium in illites to exchange. Ph.D. Diss. Iowa State University, Ames, IA. (Diss. Abstr. 67-08935).

Smith, S.J., and A.D. Scott. 1974. Exchangeability of potassium in heated fine-grained micaceous minerals. *Clays Clay Mineral.* 22:263–270.

Smith, S.J., L.J. Clark, and A.D. Scott. 1968. Exchangeability of potassium in soils. *Trans. 9th Int. Congr. Soil Sci.* 2:661–669.

Soil Conservation Service–US Department of Agriculture. 1978. Soil Survey laboratory data and descriptions for some soils of Iowa. Soil Survey Investigations Report No. 31.

Soil Science Society of America, Terminology Committees. 1979. *Glossary of Soil Science Terms*. Rev. ed. Soil Science Society of America, Madison, WI.

Soil Survey Staff. 1975. Soil taxonomy. A basic system of soil classification for making and interpreting soil surveys. US Department of Agriculture, Handbook No. 436.

Souty, N., R. Guennelon, and C. Rode. 1975. Some observations of potassium rubidium-87 and caesium-137 absorption by plants grown on nutritive solutions. *Ann. Agron.* 26:41–58.

Spencer, D.W., E.G. Degens, and G. Kulbicki. 1968. Factors affecting element distributions in sediments. In: L.H. Ahrens (ed.) *Origin and Distribution of the Elements*. Pergamon Press, Oxford, pp. 981–998.

Sreekumaran, C.K., K.C. Pillai, and T.R. Folsom. 1968. The concentrations of lithium, potassium, rubidium and cesium in some western American rivers and marine sediments. *Geochim. Cosmochim. Acta* 32:1229–1234.

Stern, K.H., and E.S. Adams. 1959. Ionic size. *Chem. Rev.* 59:1–64.

Stevens, R.E., and W.T. Schaller. 1942. The rare alkalies in micas. *Am. Mineral.* 27:525–537.

Su, N.R. 1978. Potassium in paddy soils and potassium fertilization of rice. In: *Soils and Fertilizers in Taiwan*. Taichung, Taiwan. pp. 1–22.

Suess, H.E., and H.C. Urey. 1956. Abundances of the elements. *Rev. Mod. Phys.* 28:53–74.

Swaine, D.J. 1962. The trace-element content of fertilizers. Commonwealth Bureau of Soil Science, Tech. Commun. No. 52.

Swaine, D.J., and R.L. Mitchell. 1960. Trace-element distribution in soil profiles. *J. Soil Sci.* 11:347–368.

Tabatabai, M.A. 1983. Atmospheric deposition of nutrients and pesticides. In: F.W. Schaller and G.W. Bailey (eds.) *Agricultural Management and Water Quality*. Iowa State University Press, Ames, IA, pp. 92–108.

Talibudeen, O. 1964. Natural radioactivity in soils. *Soils Fertil.* 27:347–359.

Talibudeen, O. 1972. Exchange of potassium in soils in relation to other cations. *Proc. 9th Colloq. Int. Potash Inst.* pp. 97–112.

Talibudeen, O., and M.B. Page. 1983. Ion-selective electrodes. In: K.A. Smith (ed.) *Soil Analysis*. Marcel Dekker, New York, pp. 55–113.

Tamura, T. and D.G. Jacobs. 1960. Structural implications in cesium sorption. *Health Phys.* 2:391–398.

Taylor, S.R. 1965. The application of trace element data to problems in petrology. *Phys. Chem. Earth* 6:133–213.

Tettenhorst, R. 1962. Cation migration in montmorillonites. *Am. Mineral.* 47:769–773.

Timmons, D.R., and R.F. Holt. 1977. Nutrient losses in surface runoff from a native prairie. *J. Environ. Qual.* 6:369–373.

Tinker, P.B. 1967a. A comparison of the properties of sodium and potassium in the soil. Chilean Nitrate Agriculture Service, Information No. 97.

Tinker, P.B. 1967b. The relationship of sodium in the soil to uptake of sodium by sugar beets in the greenhouse and to yield responses in the field. Trans. Commun. II and IV, International Society of Soil Science, Aberdeen, p. 223.

Tisdale, S.L., W.L. Nelson, and J.D. Beaton. 1985. *Soil Fertility and Fertilizers*. 4th ed. Macmillan, New York.

Turekian, K.K. 1969. The oceans, streams, and atmosphere. In: K.H. Wedepohl (ed.) *Handbook of Geochemistry*. Vol. 1. Springer-Verlag, Berlin, pp. 297–323.

Unger, P.W., and T.M. McCalla. 1980. Conservation tillage systems. *Adv. Agron.* 33:1–58.

Urey, H.C. 1967. The abundance of the elements with special reference to the problem of the iron abundance. *Quart. J. Roy. Astron. Soc.* 8:23–47.

US Department of Agriculture. 1979. Animal waste utilization on cropland and pastureland. USDA Utilization Res. Report No. 6. US Government Printing Office, Washington, DC.

US Salinity Laboratory Staff. 1954. Diagnosis and improvement of saline and alkali soils. US Department of Agriculture, Handbook No. 60.

Verma, G.P. 1963. Release of nonexchangeable potassium from soils and micaceous minerals during short periods of cropping in the greenhouse. Ph.D. Diss. Iowa State Univ., Ames, IA. (Diss. Abstr. 64-03997).

Vinogradov, A.P. 1959. *The Geochemistry of Rare and Dispersed Chemical Elements in Soils*. 2nd ed. Consultants Bureau, New York.

Vlasov, K.A. (ed.) 1966a. *Geochemistry and Mineralogy of Rare Elements and Genetic Types of Their Deposits*. Vol. 1. *Geochemistry of Rare Elements*. Israel Program for Scientific Translations, Jerusalem.

Vlasov, K.A. (ed.). 1966b. *Geochemistry and Mineralogy of Rare Elements and Genetic Types of their Deposits*. Vol. II. *Mineralogy of Rare Elements*. Israel Program for Scientific Translations, Jerusalem.

Weaver, C.E. 1967. Potassium, illite and the ocean. *Geochim. Cosmochim. Acta* 31:2181–2196.

Wedepohl, K.H. 1969a. Composition and abundance of common igneous rocks. In: K.H. Wedepohl (ed.) *Handbook of Geochemistry*. Vol. 1. Springer-Verlag, Berlin, pp. 227–249.

Wedepohl, K.H. 1969b. Composition and abundance of common sedimentary rocks. In: K.H. Wedepohl (ed.) *Handbook of Geochemistry*. Vol. 1. Springer-Verlag, Berlin, pp. 250–271.

Welby, C.W. 1958. Occurrence of alkali metals in some Gulf of Mexico sediments. *J. Sed. Petrol.* 28:431–452.

Welch, L.F., and A.D. Scott. 1959. Nitrification in nutrient solutions with low levels of potassium. *Can. J. Microbiol.* 5:425–430.

Welch, L.F., and A.D. Scott. 1960. Nitrification of fixed ammonium in clay minerals as affected by added potassium. *Soil Sci.* 90:79–85.

Welch, L.F., and A.D. Scott. 1961. Availability of nonexchangeable soil potassium to plants as affected by added potassium and ammonium. *Soil Sci. Soc. Am. Proc.* 25:102–104.

Wells, N., and J.S. Whitten. 1972. A pedochemical survey. Part 1. Lithium. *N.Z. J. Sci.* 15:90–106.

Wiklander, L. 1954. Forms of potassium in the soil. *Potassium Symp.* 1954:109–121.

Wiklander, L. 1964. Cation and anion exchange phenomena. In: F.E. Bear (ed.) *Chemistry of the Soil*. Reinhold, New York, pp. 163–205.

Williams, D.E. 1961. The absorption of potassium as influenced by its concentration in the nutrient medium. *Plant Soil* 15:387–399.

Xu, M.L., and Z.Y. Liu. 1982. The nutrient status of soil-root interface. I. The application of potassium selective micro-electrode. *Acta Pedol. Sinica* 19:367–374.

Xuan, J.X. 1982. Mathematical model for the movement of potassium ions to rice roots. *Acta Pedol. Sinica* 19:296–304.

# The Diagnosis and Recommendation Integrated System (DRIS)*

J.L. Walworth and M.E. Sumner

I. Introduction .......... 149
II. Nutrient Concentration and Aging .......... 150
III. DRIS Norms .......... 154
IV. Making a Diagnosis: Use of a DRIS Chart .......... 158
V. Calculating DRIS Indices .......... 160
VI. Nutrient Index Interpretation .......... 163
VII. Testing DRIS Norms .......... 166
VIII. Comparisons of DRIS and Other Diagnostic Systems .......... 170
IX. Effect of Leaf Age and Position on DRIS Indices .......... 175
X. Universality of Foliar Norms .......... 179
XI. Expansion of DRIS Beyond Nutrient Ratios .......... 182
XII. Conclusions .......... 184
References .......... 185

## I. Introduction

Foliar analysis can be a useful tool for assessing plant nutrient status only if adequate procedures are available for making diagnoses from analytical data. Because of the dynamic nature of foliar composition, which is strongly influenced by aging processes as well as interactions affecting nutrient uptake and distribution, foliar diagnosis can become a complex exercise. The diagnosis and recommendation integrated system (DRIS) was developed by Beaufils (1957, 1971, 1973) as an objective means of coping with the difficulties inherent in diagnostic procedures.

The present discussion deals initially with the effects of aging on foliar composition. The methodology of DRIS and the mechanisms by which this system minimizes these effects follow. Lastly, examples are illustrated with comparisons of diagnoses made using DRIS and other frequently used systems.

*Contribution from the Department of Agronomy, University of Georgia, Athens, GA 30602.


Advances in Soil Science, Volume 6

# II. Nutrient Concentration and Aging

It has long been recognized that nutrient concentrations (nutrient mass/ total dry plant mass) change markedly as plants age. In a detailed study of alfalfa (*Medicago sativa* L.) growth, for example, Rominger *et al.* (1975) found that N, P, K, S, Zn, Mn, and B concentrations decreased with advancing maturity (Table 1). Calcium and Mg levels, on the other hand, increased until early budding and declined thereafter. Dow and Roberts (1982) reported that nitrate, P, K, and Zn concentrations in potato (*Solanum tuberosum* L.) leaf tissue decreased as that plant aged. Similar patterns have been reported for several other crop species. Over a 10-month period, for instance, N concentration in sugarcane (*Saccarum officinarum* L.) tissue dropped from 2.70 to 1.56%, P from 0.27 to 0.16%, and K from 1.65 to 1.40% (Sumner, 1979). Corn (*Zea mays* L.) also exhibits large nutrient concentration changes during the aging process. Terman (1974) reported that N concentration in the top leaves of corn plants declined between 29 and 59% during the course of the growing season, depending on the level of N fertilization. During that period, P levels decreased between 24 and 32%, whereas K decreases were from 28 to 72%. During this same period of time, Ca and Mg concentrations increased. Similar findings were reported by Melsted *et al.* (1969) for corn plants (Figure 1) and by Sumner (1985) for peaches (*Prunus persica* Batsch) (Figure 2).

Although, as these examples illustrate, genetic variations exist, in general foliar concentrations of N, P, K, and S tend to decrease during aging. In contrast, Ca and Mg concentrations tend to increase, although declines have been reported in very early and late stages of growth in some crops.

The dynamic nature of plant tissue nutrient composition, which has been illustrated by the examples cited, imposes severe limitations on the use of foliar analysis for diagnostic purposes (Bates, 1971; Bouma, 1983). Critical nutrient value or sufficiency range systems generally depend on

**Table 1.** Changes in alfalfa composition during the growing season[a]

| Weeks | As percentage dry matter | | | | | | As mg kg$^{-1}$ dry matter | | | |
|---|---|---|---|---|---|---|---|---|---|---|
| | N | P | K | Ca | Mg | S | Mn | Zn | Cu | B |
| 1 (Vegetative) | 5.66 | 0.58 | 2.88 | 1.39 | 0.47 | 0.34 | 79.0 | 46.0 | 7.5 | 43.0 |
| 2 (Vegetative) | 4.99 | 0.46 | 2.34 | 1.57 | 0.50 | 0.35 | 76.7 | 36.5 | 7.5 | 48.5 |
| 3 (Vegetative) | 4.27 | 0.40 | 2.27 | 1.57 | 0.51 | 0.31 | 57.0 | 35.0 | 7.0 | 43.5 |
| 4 (Early bud) | 3.59 | 0.35 | 2.34 | 1.45 | 0.45 | 0.25 | 49.0 | 30.5 | 8.0 | 38.0 |
| 5 (Mid bud) | 3.15 | 0.29 | 2.04 | 1.41 | 0.43 | 0.24 | 48.0 | 23.0 | 7.5 | 37.0 |
| 6 (10% bloom) | 2.66 | 0.24 | 1.60 | 1.28 | 0.38 | 0.24 | 40.5 | 21.5 | 7.0 | 32.0 |
| 7 (60% bloom) | 2.42 | 0.23 | 1.51 | 1.25 | 0.35 | 0.24 | 41.5 | 19.5 | 7.5 | 30.5 |
| 8 (85% bloom) | 2.41 | 0.20 | 1.43 | 1.31 | 0.35 | 0.22 | 45.0 | 17.5 | 7.5 | 27.0 |
| CV (%) | 33.6 | 38.1 | 24.6 | 8.7 | 14.9 | 18.7 | 27.9 | 34.8 | 4.3 | 19.6 |

[a]Adapted from Rominger *et al.* (1975).

**Figure 1.** Effect of age on concentration of nutrients in corn leaves. Data from Table 4 (limited treatments). From Melsted *et al.* (1969).

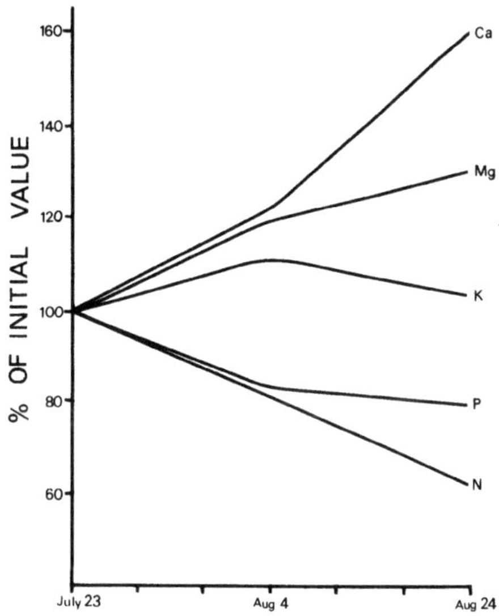


diagnostic norms derived from plant tissue of a specific age and categorize plants as healthy or unhealthy based solely on nutrient concentration (nutrient mass/dry plant mass). Thus, the stage of growth of the sampled plant is a prime concern. For this reason, diagnoses based on the use of such standards are usually applied only to plant samples harvested within narrowly specified stages of growth. For example, norms for corn tissue

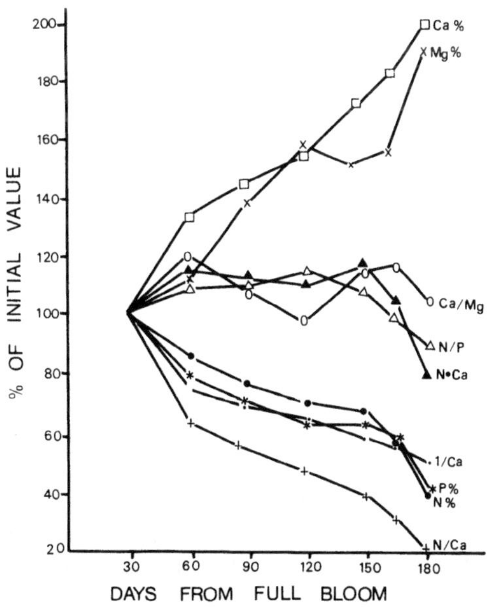


**Figure 2.** The effect of age on the different forms of expression for leaf composition of peaches. From Sumner (1985).

have generally been standardized for the ear leaf at the silking stage or for the whole plant 2 to 6 weeks after planting, restricting tissue testing to those periods (J.B. Jones and Eck, 1973).

The sampling periods for which standards have been developed often occur too late in the growing season for fertilizer applications to be effective in the treatment of nutritional problems in annual crops, or they may not coincide with the onset of visible symptoms of nutritional disorders when agriculturalists are most apt to desire diagnostic information. This situation prompted Aldrich (1973) to state that for annual agronomic and horticultural crops "plant analyses are usually postmortems."

One means of coping with the dynamic nature of foliar composition, as it pertains to diagnosis of the status of mineral nutrition in crop plants, is to develop standard tissue values for a wide range of developmental stages. These types of standards have been developed and, in fact, are available for a few of the more common crop plants (Ulrich and Hills, 1967; Munson and Nelson, 1973; Tserling, 1974). One can then sample plants through a range of growth stages and apply the corresponding critical nutrient values or sufficiency ranges. Although simple in theory, this procedure can be difficult in practice. First, an accurate determination of the stage of growth must be made in the field at the time of sampling. Thus, the accuracy of this system is dependent upon a subjective determination by a sampler who may not be trained or experienced in such procedures. Second, the sampler must accurately communicate this information to the diagnostician so the proper norms can be selected. There may be further difficulties in the implementation of this system. Little research has been initiated to determine the influence of cultivar maturity group on nutrient concentrations at given growth stages. Also, the effects of conditions that delay or accelerate maturation processes of plants may influence the relationship between nutrient concentration and developmental stage. The relative effects of time and growth conditions (light, temperature, etc.) on chemical composition of plants have not, as yet, been described adequately to insure accurate diagnoses using this process.

An alternative to this approach was suggested as a part of DRIS by Beaufils (1973), who reasoned that if N, P, and K concentrations (relative to dry matter) all decrease during aging, then the ratios N/P, N/K, and P/K (or their reciprocals) should remain fairly constant. Similarly, because Ca and Mg concentrations usually increase during aging, a quotient formed from these two nutrients (Ca/Mg or Mg/Ca) should also produce a constant value. Furthermore, products of two nutrients, one of whose content relative to dry matter decreases and the other increases with time, ($N \times Ca$, for example) should also be fairly constant. These concepts have proved correct within moderate ranges of age and are illustrated in Figure 2 (Sumner, 1985). Nitrogen and P contents of peaches, when presented as ratios with dry matter (this form is usually just called "percent," the denominator, dry matter, being assumed and left unstated) change rapidly,

**Table 2.** Coefficients of determination ($r^2$) between age of corn crop at sampling and form of expressing tissue composition[a]

| Form of expression | Coefficient of determination, $r^2$ |
|---|---|
| Dry matter basis (%) | |
| N | 0.41 |
| P | .25 |
| K | 0.55 |
| Ratio basis | |
| N/P | 0.00 |
| N/K | 0.06 |
| K/P | 0.01 |

[a]Data from Beaufils (1971).

dropping to less than 50% of their original values over 5 months. Yet the N/P ratio, although fluctuating slightly, was still 90% of the initial value after 5 months. Over the same period of time, Ca and Mg concentrations increased about 90 and 100%, respectively, while the Ca/Mg ratio remained essentially unaltered. The N × Ca product was similarly constant over the 5-month period of this study, demonstrating the usefulness of product expressions when relating nutrients exhibiting opposite behavior (increasing versus decreasing nutrient concentrations).

A further example of the use of nutrient ratios is illustrated in Table 2, where corn tissue compositions were expressed both as ratios with dry matter and with other nutrients. The dry matter forms of expression (N%, P%, and K%) were much more closely related to age than were expressions in the form of nutrient ratios. Similarly, coefficients of variation for the alfalfa nutrient ratios presented in Table 3 were much smaller than were

**Table 3.** Effect of age on the coefficient of variation (CV) of several nutrient expressions in alfalfa sampled over an 8-week period[a]

| Nutrient Expression | CV (%) |
|---|---|
| N% | 33.6 |
| P% | 38.1 |
| K% | 24.6 |
| N/P | 6.2 |
| N/K | 12.5 |
| P/K | 15.0 |

[a]Calculated from data of Rominger *et al.* (1975).

those for N, P, and K on a dry plant mass basis, indicating that by calculating ratios the influence of age is reduced.

The constancy of nutrient ratios and products, as illustrated in the previous examples, is not without limits. Rates of change of concentrations of several nutrients may be quite rapid in very young plants and nutrient ratios and products may be correspondingly variable at this time. Nevertheless, these forms of expression are often less affected by aging processes and so present an opportunity to expand the usefulness and accuracy of foliar diagnoses.

Nutrient ratio expressions require no extra analytical measurements because

$$N\%/P\% = (100N/DM)/(100P/DM) = (100N/DM) \times (DM/100P) = N/P.$$

where DM is dry matter. Even though a ratio such as N/P can be easily calculated from N% and P%, commonly reported values, interpretation may be more difficult (Sumner, 1978). For example, if the optimum corn leaf N/P ratio is 10.04 and an individual sample has a value of 14.00, it is impossible to determine whether N is too high, P is too low, or N is too high *and* P is too low.

Parenthetically, nutrient percentages are also ratios, N%, for example, being composed of mass of N in the numerator and dry matter mass in the denominator. Therefore, if N% is too high in a plant sample, one cannot determine whether N is too high or dry matter is too low, although the former is usually assumed because dry matter is often treated as a constant (which, of course, it is not!). This condition could result from a situation where dry matter accumulation was limited while N uptake was not and could be interpreted either as dry matter deficiency or N excess.

Taking a holistic approach and considering a series of ratios for a single nutrient (N/P, N/K, etc.) is even more confusing than considering a single ratio and is probably one reason that dry matter ratios (N%, P%, etc.) have been preferred over nutrient ratios. In essence, DRIS provides a means of ordering nutrient ratios into meaningful expressions, which are called DRIS indices. Simply stated, a nutrient index is a mean of the deviations of the ratios containing a given nutrient from their respective normal or optimum values. The nutrient indices, then, give an indication of the relative status of nutrient constituents in plant material relative to other constituents included in the diagnosis. In practice, the system is slightly more complex than this description indicates, but keeping this general model in mind will make the following procedural description more easy to comprehend.

## III. DRIS Norms

The first step in implementing DRIS or any other foliar diagnostic system is the establishment of standard values or norms. The DRIS utilizes a survey approach (Beaufils, 1971, 1973) for norm determination that is based on a crop response model similar to that illustrated in Figure 3 (Sumner and Farina, 1986). In this model, only the uppermost line represents conditions under which yield is actually limited by the variable plotted on the abscissa. The peak of this line represents the optimum value for the yield-determining factor plotted on the abscissa. The lower lines in Figure 3 represent yield responses under suboptimal conditions, or when other variables limit yield. Data comprising the uppermost curve in this model cannot be obtained using a single factor experiment because optimal levels of other variables continually change via various interactions. Therefore, a survey approach is desirable and has been generally adopted for norm determination in DRIS.

An actual data set, collected from around the world, and comprised of corn yield and ear leaf N concentrations is plotted in Figure 4, along with an empirically derived delimiter or boundary line. This boundary line represents the limits of corn yield when tissue N is the sole yield-determining factor and is analagous to the generalized limit represented by the uppermost line in Figure 3. The top portion of this boundary line may be left open to indicate an undefined genetic yield potential.

For a normal or Gaussian distribution of observations, the mean of all values provides an accurate estimate of the peak or maximum value (Steel and Torrie, 1980). Unfortunately, few populations of foliar nutrient data collected in the field are normally distributed. A visual inspection of the scatter of points in Figure 4 reveals a bias in which low yields are more frequently associated with low than with high tissue N. This should not be surprising because, although excessive N can certainly induce low yields, farmers producing low yields are more likely to under- than to

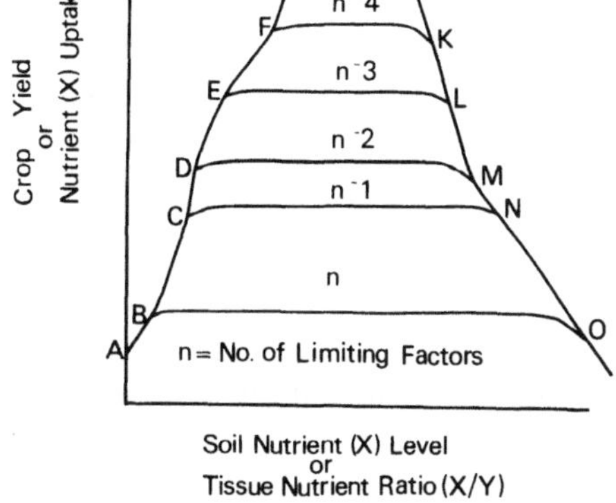


**Figure 3.** Diagrammatic representation of crop response to a number of limiting factors. From Sumner and Farina (1986).

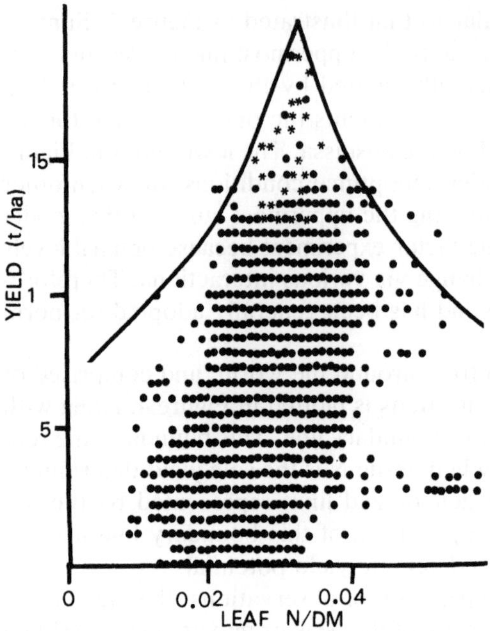


**Figure 4.** Maize yield versus leaf N/DM (percent per 100) including over 8000 data points collected worldwide. Included is a boundary line confining the data. From Walworth *et al.* (1986a).

overfertilize. Therefore, the poor distribution in Figure 4 may be more indicative of agricultural practices than of plant responses. Nevertheless, skewed distributions can represent a problem if such data are to be used for standard value generation.

In DRIS, this difficulty has been overcome by dividing data sets or populations of observations into high- and low-yielding subgroups, then averaging values only from the high-yielding group to obtain estimates of tissue parameter optima. In addition, the coefficients of variation (CVs) of the high-yielding data provide a measure of the relative spread or breadth of the yield response surface at upper yield levels.

The actual cutoff value used to divide high- and low-yield groups is not critical as long as the high-yield data remain normally distributed. When the cutoff value for dividing high and low corn yields was varied from 7 to 9 Mg $ha^{-1}$, the normal values for tissue N/P, N/K, and P/K ratios determined by averaging varied only 6.7, 4.8, and 2.4%, respectively (Letzsch and Sumner, 1984). In practice, chosen cutoff values usually represent yields that "better" farmers routinely obtain.

For each pair of nutrients, there are three forms of expression that may be considered. Nitrogen and P, for example, can be related as the ratio N/P, as the inverse P/N, or as the product N $\times$ P. In DRIS calculations only one expression is used to relate each nutrient pair, and the selection is made in the following manner. Consider the yield versus nutrient expression distribution represented schematically in Figure 5. If a point

**Figure 5.** Schematic illustration of DRIS norm derivation. Shaded areas represent the only members that can be determined to be unhealthy because of the insufficiency, excess, or imbalance of the nutrients in question. From Walworth and Sumner (1986).

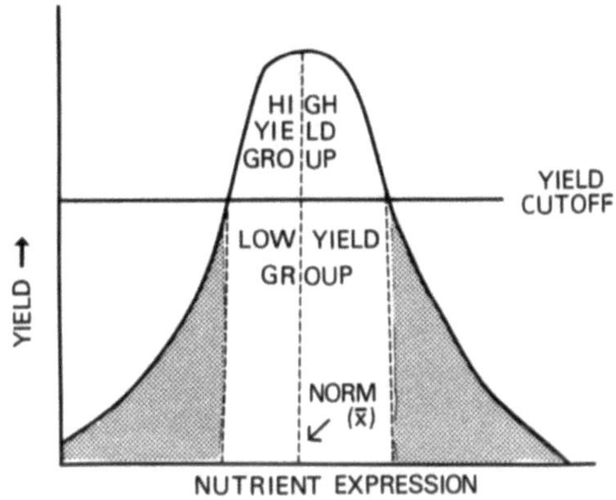


is located somewhere within this population and only the value of the nutrient expression is known, the following determinations can be made. If the nutrient value falls within the unshaded portion of Figure 5, then the individual in question may belong to either the high- or the low-yielding group. In other words, there is no way to discriminate between a healthy or an unhealthy plant. Conversly, if the value of the nutrient expression locates an individual within the shaded area, it unequivocally belongs to the low-yielding or unhealthy population. Therefore, it is desirable to maximize the shaded area in the expression selection process in order to increase diagnostic sensitivity. This is accomplished by comparing the variance of the low-yielding group to that of the high-yielding segment of the population. The form of expression (N/P, P/N, or N × P) selected for use within DRIS calculations is that with the largest variance ratio.

The origin of the data used for nutrient expression selection may, of course, profoundly affect the outcome of this process. If one expects a selected form of nutrient expression to exhibit stability with respect to aging, the data base must represent tissue values reflecting a range of ages. For example, the hypothetical nutrient ratio $X/Y$ may have a large variance ratio in a population of tissue data selected from conditions of uniform plant age. If, however, $X$/DM increases with age while $Y$/DM declines, the ratio $X/Y$ will change rapidly and may be a poor parameter selection. If, on the other hand, the data base used for expression selection were comprised of samples selected from throughout the growing season, $X/Y$ would presumably have a lower variance ratio than $X \times Y$, which would remain more constant over time. The product form would then be selected as the preferred form of expression relating these two nutrients. Thus, the data base upon which DRIS expressions are based should represent the conditions to which the system is ultimately to be applied.

To date, nearly all sampling for the purpose of DRIS norm generation has been carried out under standardized conditions (specific plant parts at specific growth stages), partially out of convenience, and partially be-

**Table 4.** Corn leaf norms for N, P,and K[a]

| Form of expression[b] | Low-yielding population (A) Mean | CV (%) | Variance $S_A$ | High-yielding population (B) Mean | CV (%) | Variance $S_B$ | Variance ratio $S_A/S_B$ |
|---|---|---|---|---|---|---|---|
| N (% DM) | 2.86 | 20 | 0.326 | 3.06 | 18 | 0.303 | 1.075 |
| P (% DM) | 0.30 | 20 | 0.0036 | 0.32 | 22 | 0.0050 | 0.720 |
| K (% DM) | 2.32 | 27 | 0.392 | 2.12 | 23 | 0.238 | 1.647[c] |
| N/P | 9.88 | 18 | 3.158 | 10.04 | 14 | 1.996 | 1.582[c] |
| N/K | 1.39 | 28 | 0.150 | 1.49 | 21 | 0.101 | 1.485[c] |
| K/P | 6.94 | 29 | 4.000 | 6.74 | 22 | 2.222 | 1.800[c] |
| P/K | 0.13 | 26 | 0.0011 | 0.15 | 24 | 0.0013 | 0.846 |
| P/N | 0.10 | 18 | 0.00032 | 0.10 | 16 | 0.00026 | 1.231 |
| K/N | 0.81 | 24 | 0.0380 | 0.72 | 22 | 0.0259 | 1.467[c] |
| NP | 0.85 | 33 | 0.0792 | 0.98 | 32 | 0.0961 | 0.824 |
| NK | 6.59 | 34 | 5.040 | 6.45 | 34 | 4.910 | 1.026 |
| PK | 0.71 | 37 | 0.0675 | 0.68 | 36 | 0.0611 | 1.105 |

[a]From Sumner (1982).

[b]DM = dry matter.

[c]Variances of low-yielding and high-yielding populations are significantly different at the 1% level.

cause sampling is often done as part of research conducted primarily for other purposes.

An example of DRIS norms derived as illustrated in Figure 5 are presented in Table 4. The forms of expression found best to discriminate between the high- and low-yielding subpopulations have been selected on the basis of the variance ratio and are N/P, N/K, and K/P.

# IV. Making a Diagnosis: Use of a DRIS Chart

In the simplest case the norms of three selected nutrients can be related to one another in a so-called DRIS chart (Figure 6). The point of intersection of the three axes corresponds to the mean value for the high-yielding population for each form of expression (N/P = 10.04, N/K = 1.49, K/P = 6.74). This is the composition desired in order to increase the chances of obtaining a high yield. However, this desired composition should be considered not a single inflexible point but a range encompassed by the inner of the two concentric circles in Figure 6. The diameter of this circle is set as 4SD/3 (Beaufils, 1971) where SD is the standard deviation of the high-yielding subpopulation. A plant composition falling

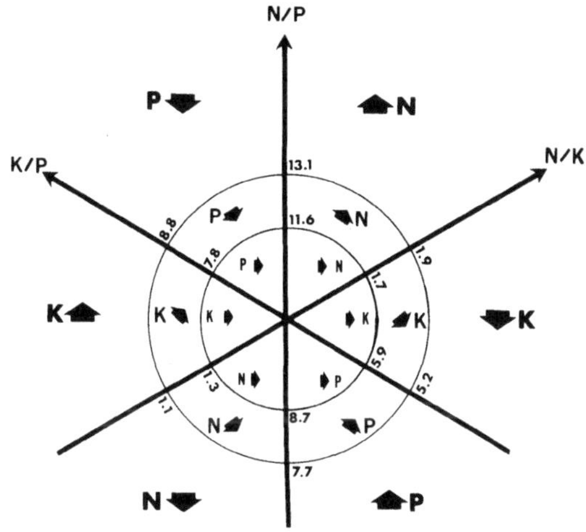


**Figure 6.** DRIS chart for obtaining qualitative order of requirement for N, P, K in corn. From Sumner (1982).

within the inner circle would be considered to be balanced and is denoted by a horizontal arrow (→). As one moves away from the central zone along any axis the degree of imbalance between the two elements increases. This zone of imbalance is divided into two subzones, the first being a zone of slight to moderate imbalance. This is denoted by an arrow at 45° to the horizontal (↘) (↗) and is encompassed by the outer of the concentric circles, which has a diameter of 8SD/3. Beyond this circle is a zone of marked imbalance denoted by vertical arrows (↑) (↓).

The way in which this chart can be used to make a diagnosis may be illustrated by means of an example. Assume that the N, P, and K concentrations in a corn leaf sample on a dry matter basis are 3.30, 0.200, and 1.20%, respectively, which give N/P = 16.5, N/K = 2.75, and K/P = 6.00. Because an excess of one nutrient corresponds to a shortage of another in terms of balance, only insufficiencies are recorded by convention for the purpose of diagnosis, which is done stepwise for each function. Identical diagnoses are obtained by considering either excesses or insufficiencies or both. Using the above data, the value of the function N/P lies beyond the outer circle in the zone of P insufficiency giving: (1) N P ↓ K, whereas the value of N/K also lies outside the outer circle in the zone of K insufficiency, giving (2) N P ↓ K ↓ and that of K/P lies inside the inner circles in the zone of balance: (3) N P ↓ K ↓. Once the three common functions have been read the remaining element is assigned a horizontal arrow. The final reading then becomes (4) N → P ↓ K ↓, which give the order of requirement for N, P, and K in terms of limiting importance on yield as: K = P > N. This does not necessarily mean that K and P are deficient and N excessive; instead, it should be considered a relative ranking of the nutrients. Without information on the status of the other growth factors, it is not possible to determine whether K and

P, which were found to be more limiting than N, are in fact determining yield. Thus, consideration of a greater number of nutrients in the diagnosis improves confidence that those nutrients relatively in short supply may be determining yield, if nutrition happens to be among the most limiting growth factors. As the plant is a dynamic system, there is no reference point to which this can be referred on an absolute basis.

## V. Calculating DRIS Indices

Although the use of this diagram enables one to make three nutrient diagnoses, DRIS also provides a mathematical means of ordering a large number of nutrient ratios and/or products into nutrient indices that can be easily interpreted. Initially, standard values or norms must be determined, in the manner described, for all nutrient ratios or products (i.e., all two-way pairings of nutrients) to be used in index calculations. The standard values are then used to generate indices by the following equations, in this case for the hypothetical nutrients A through N:

$$\text{A index} = \frac{[f(\text{A/B}) + f(\text{A/C}) + f(\text{A/D}) \ldots + f(\text{A/N})]}{z} \qquad [1]$$

$$\text{B index} = \frac{[-f(\text{A/B}) + f(\text{B/C}) + f(\text{B/D}) \ldots + f(\text{B/N})]}{z} \qquad [2]$$

$$\text{N index} = \frac{[-f(\text{A/N}) - f(\text{B/N}) - f(\text{C/N}) \ldots -f(\text{M/N})]}{z} \qquad [3]$$

where, when $\text{A/B} \geqslant a/b$,

$$f(\text{A/B}) = \left(\frac{\text{A/B}}{a/b} - 1\right)\frac{1000}{\text{CV}} \qquad [4]$$

or, when $\text{A/B} < a/b$,

$$f(\text{A/B}) = \left(1 - \frac{a/b}{\text{A/B}}\right)\frac{1000}{\text{CV}} \qquad [5]$$

in which A/B is the value of the ratio of the two elements in the tissue of the plant being diagnosed, $a/b$ is the optimum value or norm for that ratio, CV is the coefficient of variation associated with the norm, and $z$ is the number of functions comprising the nutrient index. Values for the other functions, such as $f(\text{A/C})$, $f(\text{A/D})$, etc., are calculated in the same way as $f(\text{A/B})$, using the appropriate norms and CVs.

A nutrient index, then, is simply a mean of functions of all ratios containing a given nutrient. The components of this mean are weighted by the reciprocals of the CVs of the high-yielding populations from which the ratio optima or norms are developed. Therefore, if the expressions

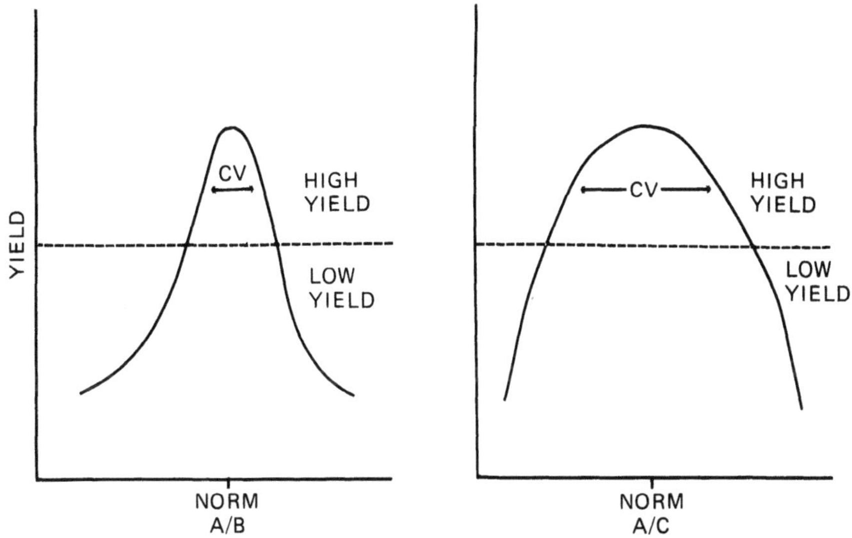


**Figure 7.** Schematic representations of relationships between nutrient expressions and yield.

A/B and A/C both are used to generate an index for the nutrient A, their contribution to the index would depend on the CVs associated with their optima, which reflect the relative influence of these two expressions on crop yield. This is illustrated in Figure 7, where the CV of the high-yielding population of A/B is much smaller than that of A/C. This indicates that, for a plant to be healthy, A/B must be relatively closer to its optimum value than must A/C. In other words, A/B is more critically related to yield than is A/C. Because A/B has the smaller CV, and each function is weighted by the reciprocal of its respective CV, A/B will contribute more to the A index than will A/C.

Each "function" is a comparison of the ratio found in an individual plant sample with the norm for that ratio. For example,

$$f(\mathrm{A/B}) = \left(\frac{\mathrm{A/B}}{a/b} - 1\right)\frac{1000}{\mathrm{CV}} \qquad [4]$$

when $\mathrm{A/B} \geq a/b$. The sample value (A/B) is compared to the norm value ($a/b$) by division. If the balance of A/B is optimum in a plant sample, then $(\mathrm{A/B})/(a/b)$ equals 1; thus $[(\mathrm{A/B})/(a/b)] - 1$, and consequently $f(\mathrm{A/B})$, equal 0. Alternatively, if A/B is greater than the norm ($a/b$), then $[(\mathrm{A/B})/(a/b)] - 1$ and $f(\mathrm{A/B})$ are positive numbers. Finally, if A/B is less than the norm, the equation used to calculate $f(\mathrm{A/B})$ is:

$$\left(1 - \frac{a/b}{\mathrm{A/B}}\right)\frac{1000}{\mathrm{CV}} \qquad [5]$$

Therefore 1 − [(*a*/*b*)/(A/B)] and *f*(A/B) are negative numbers. In either case the 1000 multiplier in the function equations is 100 × 10, with the value 10 being included as a matter of practicality to give the resultant indices convenient magnitudes and having no actual functional purpose. The value 100 is the denominator of the coefficient of variation expressed as a percentage.

The need for two separate function equations, dependent on whether the sample ratio is smaller than or larger than the norm, can best be explained in the following manner. The function of A/B is used in the calculation of indices for both nutrients A and B. When calculating the A index, *f*(A/B) is added to the other functions prior to averaging {index A = [*f*(A/B) + *f*(A/C) . . . ]/z}. However, this same function is used with a reversed sign in calculations of the index for B {index B = [−*f*(A/B) + *f*(B/C) . . . ]/*z*}.

The value of the ratio A/B contains information pertaining to the relative values of A and B, and, as mentioned earlier, if A/B is nonoptimal, it cannot be determined whether A and/or B are too large or too small, or some combination of the two. Because A/B is a measure of A relative to B or B relative to A, this ratio should contribute equally to the A and B indices. If, for example, the optimum value for this ratio [*a*/*b* in the *f*(A/B) equation] were equal to 1 and the actual value from a plant sample A/B in f(A/B) were 0.75, then, when A/B < *a*/*b*,

$$f(\mathrm{A/B}) = \left(1 - \frac{a/b}{\mathrm{A/B}}\right)\frac{1000}{\mathrm{CV}}$$

$$= \left(1 - \frac{1}{0.75}\right)\frac{1000}{\mathrm{CV}}$$

$$= -0.33\ (1000/\mathrm{CV})$$

However, if the sample value of A/B is 1.33 (1/0.75) then, when A/B < *a*/*b*,

$$f(\mathrm{A/B}) = \left(\frac{\mathrm{A/B}}{a/b} - 1\right)\frac{1000}{\mathrm{CV}}$$

$$= \left(\frac{1.33}{1} - 1\right)\frac{1000}{\mathrm{CV}}$$

$$= 0.33\ (1000/\mathrm{CV})$$

Therefore, the contribution from A and B are equal in the nutrient index expressions.

If this correction were not made, the contribution of the numerator and denominator would be unequal. If, for example, the sample value of A/B were 1.33 and the wrong equation were used,

$$f(\mathrm{A/B}) = \left(1 - \frac{a/b}{\mathrm{A/B}}\right)\frac{1000}{\mathrm{CV}}$$
$$= \left(1 - \frac{1}{1.33}\right)\frac{1000}{\mathrm{CV}}$$
$$= 0.25\ (1000/\mathrm{CV})$$

rather than $f(\mathrm{A/B}) = 0.33\ (1000/\mathrm{CV})$ obtained with the correct function equation.

The impact of nutrient changes on DRIS indices then would be to some degree a function of their position in nutrient expressions, i.e., whether they appear in the numerator or the denominator of a nutrient ratio. However, because two different function equations are used, inverse ratios, when introduced into the function equations [$f$(A/B), etc.] result in additive rather than multiplicative inverses, and calculated indices are independent of the form of expression used.

C.A. Jones (1981) made the proposal of using equation [4] in all cases to calculate the value $f$(A/B) as he claimed that equation [5] systematically overestimates the deviation of A/B from $a/b$ when A/B $< a/b$. This is not true, as can be seen from the foregoing discussion. Elwali and Gascho (1983, 1984) also proposed a modification in the calculation of indices, which consists of considering two nutrients to be in balance if the ratio of their concentrations in the sample (A/B) was within the range given by the general mean plus or minus the SD of that ratio ($a/b \pm \mathrm{SD}_{a/b}$).

## VI. Nutrient Index Interpretation

Because the value of each ratio function is added to one index subtotal and subtracted from another [i.e., the quantity $f$(A/B) is added to the A index subtotal and subtracted from that of the B index] prior to averaging, all indices are balanced around zero. Therefore, nutrient indices should sum to zero. The more negative an index, the more lacking is the nutrient it represents relative to other nutrients used in the diagnosis. Alternatively, a large positive nutrient index indicates that the corresponding nutrient is present in relatively excessive quantity.

An example will serve to illustrate. For simplicity, this example is limited to the nutrients N, P, and K. Other nutrients may be incorporated in an identical manner. The norms currently being used for corn ear leaf diagnoses and their associated CVs, are presented in Table 4. If a corn ear leaf sample contained 3.30% N, 0.20% P, and 1.20% K on a dry matter basis, the calculations would proceed as follows. The sample ratios (represented in capital letters in the previous equations) are N/P = 3.30/0.20 = 16.5, N/K = 3.30/1.20 = 2.75 and K/P = 1.20/0.20 = 6.00. Therefore,

$$f(\mathrm{N/P}) = \left(\frac{\mathrm{N/P}}{n/p} - 1\right)\frac{1000}{\mathrm{CV}}$$

because N/P > $n/p$. Inserting the proper values, this equals [(16.5/10.04) − 1] (1000/14) = 45.96.

Similarly

$$f(\mathrm{N/K}) = \left(\frac{\mathrm{N/K}}{n/k} - 1\right)\frac{1000}{\mathrm{CV}} = \left(\frac{2.75}{1.49} - 1\right)\frac{1000}{\mathrm{CV}} = 40.27$$

The equation for $f$(K/P), however, is 1 − [($k/p$)/(K/P)](1000/CV) because $k/p$ > K/P and becomes [1 − (6.74/6.00)](1000/CV) = −5.61. The nutrient indices are then calculated:

$$\text{N index} = [f(\mathrm{N/P}) + f(\mathrm{N/K})]/2 = (45.96 + 40.27)/2 = 43$$

$$\text{P index} = [-f(\mathrm{N/P}) - f(\mathrm{K/P})]/2 = (-45.96 + 5.61)/2 = -20$$

$$\text{K index} = [-f(\mathrm{N/K}) + f(\mathrm{K/P})]/2 = (-40.27 - 5.61)/2 = -23$$

Therefore, N index (+43) >> P index (−20) > K index (−23). This is interpreted to mean that in terms of relative importance to yield, K is more required than P, which is much more required than N. This is essentially the same diagnosis as was found using arrow notation (see Section IV). Through the use of a numerical diagnosis, one can further conclude that K is slightly more insufficient in the sampled plant than is P.

In a plant sample with optimal nutrient balance, all nutrient indices would equal zero. However, it is important to recognize that an individual nutrient is not necessarily present in optimum concentration if its index equals zero. If, for instance, results of a diagnosis were as follows:

| Nutrient | N | P | K | Ca | Mg |
|---|---|---|---|---|---|
| Index | −21 | 0 | +7 | +7 | +7 |

one could accurately say that, of all nutrients tested, N had the most negative index, was relatively least abundant, and was likely to be yield limiting if nutrition were governing growth. Although the P index equaled 0, it was relatively less abundant than K, Ca, or Mg and was the second most needed nutrient in this diagnosis. Potassium, Ca, and Mg levels were excessive relative to N and P. In this example K, Ca, and Mg may have actually been more yield limiting than P, for instance. However, because nutrients can, in practical terms, be added but not taken away, the recommendations from this diagnosis include supplementing the supply of N and, to a lesser extent P, even though the P index equals 0.

Some measure of the total nutritional balance in a plant may be indicated by the sum of the nutrient indices irrespective of sign. This is illustrated by Figure 8, the data for which are taken from Meyer (1975) and are comprised of mean sugarcane yields and sums of nutrient index values irre-

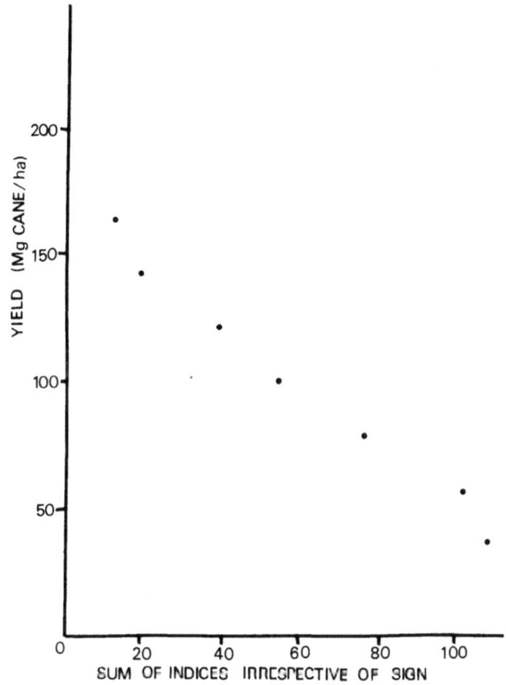


**Figure 8.** Relationship between sum of DRIS indices irrespective of sign and yield of sugarcane. Data from Meyer (1975, *Table 2*).

spective of sign from factorially arranged NPK field experiments. The relationship between nutrient balance and yield is immediately evident, the yield generally decreasing with increasing sum of indices. In general, however, yield cannot be predicted from sum of indices irrespective of sign because of the influence of unmeasured factors that may affect yield but not calculated DRIS indices. The relationship in Figure 9, constructed of data compiled from several experiments where other nutrients and environmental conditions were allowed to vary, reflects the effects that the

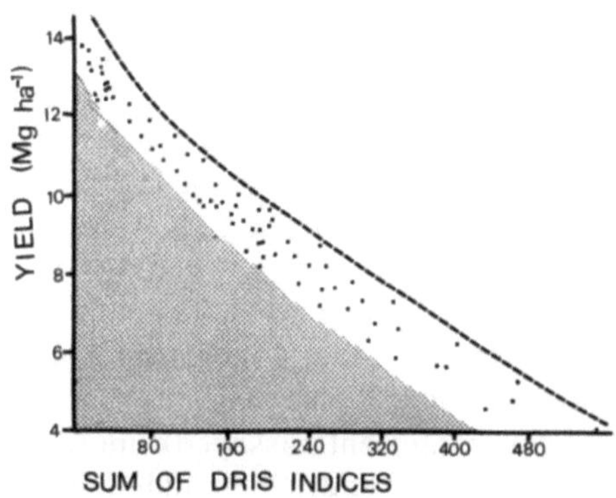


**Figure 9.** Scatter diagram for relationship between corn yield and sum of DRIS indices irrespective of sign. From Sumner (1977a).

unmeasured parameters have on yield. When the sums of the DRIS indices are large, one or more of the measured factors (N, P, and K in this case) limit yields. Consequently, a large yield cannot occur when the sum of indices is large. Large yields can result only when the sum of indices is small (when N, P, and K are balanced), although low yields may still occur if other nutritional or environmental factors are limiting. This is illustrated by the shaded portion of lower yielding observations. Presumably, if indices were calculated for all yield-determining factors, yield would be proportional to, and could be calculated from, the sum of DRIS indices in a given situation. The diagnosis and recommendation integrated system can be envisioned as a modified regression technique that uses boundary line conditions of an incomplete set of independent variables to describe the dependent variable, yield. The result, calculated as the sum of indices irrespective of sign, is indicative of maximum attainable yield, but additional information (weather conditions, disease, other nutrients, insects, planting date, plant population, weeds, etc.) would be needed to complete the equation and actually predict yield.

## VII. Testing DRIS Norms

Diagnosis and recommendation integrated system norms, calculated on the basis of finite sets of field data, must be tested to insure validity and accuracy. To do this, DRIS diagnoses are usually conducted on field- or greenhouse-grown plants selected from factorially designed fertilizer trials. It is imperative that these data be independent from those used to generate the norms and CVs used in index calculations. The following procedure may be used. First, using data from an experiment in which yield responses have been obtained to the nutrients being studied, plants from the control or lowest treatment level are diagnosed, and the most needed nutrient(s) determined. Then the treatment with additions prescribed by the initial diagnosis is located and the yields are compared. If yield increases when the appropriate treatment is applied, then that diagnosis is considered a success; if not, it is considered a failure. One then proceeds with an evaluation of the nutritional status of that second treatment and so forth, until all indices equal zero or, more commonly, until the prescribed treatment cannot be found as part of the experimental layout.

Examples of this type of progressive diagnosis are presented in Tables 5, 6, and 7. In Table 5, tissue composition and corresponding DRIS indices for alfalfa samples from a PKS factorial experiment are presented. In the base treatment ($P_0K_0S_0$), K had the lowest DRIS index and so is considered the most lacking of the seven elements in the tissue assay. When K was added ($P_0K_1S_0$), the K index increased as did the yield, while the sum of indices declined. This indicates that the overall nutritional balance (among the seven nutrients assayed) improved. However, K still had the most negative index and, when added at the next level ($P_0K_2S_0$), resulted in a

**Table 5.** Progressive diagnosis of alfalfa tissue with corresponding nutrient concentrations and yields[a]

| Treatment | | | Plant composition | | | | | | | DRIS indices | | | | | | | Sum of indices irrespective of sign | Total dry matter yield |
|---|---|---|---|---|---|---|---|---|---|---|---|---|---|---|---|---|---|---|
| P | K | S | (N) (%) | P (%) | K (%) | S (%) | Ca (%) | Mg (%) | B (ppm) | N | P | K | S | Ca | Mg | B | | (Mg ha$^{-1}$) |
| 0 | 0 | 0 | 3.49 | 0.22 | 1.05 | 0.29 | 1.70 | 0.47 | 42 | 12 | −19 | −59 | 15 | 12 | 24 | 16 | 156 | 9.83 |
| 0 | 1 | 0 | 3.37 | 0.23 | 1.61 | 0.30 | 1.62 | 0.38 | 47 | 3 | −20 | −25 | 11 | 6 | 9 | 16 | 91 | 11.53 |
| 0 | 2 | 0 | 3.34 | 0.20 | 1.97 | 0.28 | 1.39 | 0.32 | 39 | 6 | −24 | −10 | 10 | 3 | 5 | 11 | 69 | 11.55 |
| 1 | 2 | 0 | 3.44 | 0.23 | 1.68 | 0.26 | 1.41 | 0.33 | 49 | 7 | −17 | −21 | 4 | 2 | 5 | 20 | 75 | 11.66 |
| 1 | 3 | 0 | 3.33 | 0.24 | 2.51 | 0.28 | 1.38 | 0.30 | 53 | 0 | −18 | −3 | 4 | −2 | −1 | 20 | 48 | 12.27 |
| 2 | 3 | 0 | 3.29 | 0.26 | 2.60 | 0.25 | 1.33 | 0.28 | 49 | 0 | −11 | −1 | −1 | −2 | −2 | 17 | 35 | 12.50 |

[a]From Erickson *et al.* (1982).

further decrease in the sum of indices, although essentially no increase in yield. This may have been caused by a shortage of P resulting from the added K, for the P index was the most negative in the $P_0K_2S_0$ plot. Adding P ($P_1K_2S_0$) as suggested by this diagnosis, resulted in a further yield increase and a decline of the index sum, although it also induced a K shortage. The next K addition ($P_1K_3S_0$) increased yield and improved nutritional balance but again induced a P insufficiency. Therefore, additional P was the proper treatment ($P_2K_3S_0$) and the result was once again improved nutritional balance and increased yield.

The chances of making successful recommendations based on foliar diagnoses are increased as the number of yield-affecting factors considered is increased. Tables 6 and 7 serve to illustrate this point. In Table 6, progressive diagnoses have been performed on corn leaves selected from a $3^4$ NPKS factorial experiment using DRIS indices calculated initially only from foliar N, P, and K data. In the control, ($N_1P_1K_1S_1$), the most needed nutrient appeared to be N, which, when added ($N_2P_1K_1S_1$), produced a yield increase. Phosphorus then had the most negative index. Upon addition of P ($N_2P_2K_1S_1$), the yield further increased, with K becoming the most required index. On addition of K ($N_2P_2K_2S_1$), yield increased further and N was diagnosed as most limiting. However, addition of N ($N_3P_2K_2S_1$) resulted in a substantial yield decline. Phosphorus was then diagnosed as most needed, and when added ($N_3P_3K_2S_1$), yield again reached the level in $N_2P_2K_2S_1$. It therefore appears that DRIS has failed to diagnose properly the nutritional status of plants at the $N_2P_2K_2S_1$ level of this experiment.

**Table 6.** Progressive diagnosis of N, P, and K requirements of corn in the presence and absence of S, which is required by the plant [a]

| Treatment | | | | Leaf composition (%) | | | | DRIS indices | | | Dry matter |
|---|---|---|---|---|---|---|---|---|---|---|---|
| N | P | K | S | N | P | K | S | N | P | K | yield (g/pot) |
| 1 | 1 | 1 | 1 | 1.44 | 0.149 | 1.37 | 0.11 | −7 | −5 | 12 | 5.04 |
| 2 | 1 | 1 | 1 | 3.40 | 0.158 | 1.80 | 0.03 | 21 | −29 | 8 | 5.59 |
| 2 | 2 | 1 | 1 | 1.82 | 0.307 | 0.83 | 0.07 | −6 | 27 | −21 | 8.66 |
| 2 | 2 | 2 | 1 | 1.96 | 0.280 | 3.93 | 0.06 | −33 | −10 | 42 | 9.59 |
| 3 | 2 | 2 | 1 | 2.74 | 0.238 | 3.20 | 0.06 | −8 | −17 | 25 | 6.56 |
| 3 | 3 | 2 | 1 | 2.82 | 0.364 | 3.47 | 0.07 | −15 | −3 | 19 | 9.99 |
| 1 | 1 | 1 | 3 | 1.56 | 0.149 | 1.79 | 0.20 | −9 | −12 | 22 | 5.02 |
| 1 | 2 | 1 | 3 | 1.08 | 0.267 | 0.91 | 0.24 | −25 | 32 | −7 | 7.88 |
| 2 | 2 | 1 | 3 | 2.78 | 0.277 | 0.87 | 0.21 | 10 | 13 | −23 | 10.59 |
| 2 | 2 | 2 | 3 | 2.02 | 0.200 | 3.38 | 0.15 | −20 | −22 | 4 | 15.56 |
| 2 | 3 | 2 | 3 | 2.76 | 0.269 | 3.34 | 0.18 | −11 | −13 | 26 | 17.06 |
| LSD (0.05) | | | | | | | | | | | 0.60 |
| LSD (0.01) | | | | | | | | | | | 0.82 |

[a] From Sumner, (1981).

**Table 7.** Progressive diagnosis of N. P, K, Ca, Mg, and S requirements of corn using DRIS indices on data from a $3^4$ factorial experiment with N, P, K, and S as factors[a]

| Treatment | | | | Leaf composition (%) | | | | | | DRIS indices | | | | | | Dry matter |
|---|---|---|---|---|---|---|---|---|---|---|---|---|---|---|---|---|
| N | P | K | S | N | P | K | Ca | Mg | S | N | P | K | Ca | Mg | S | yield (g/pot) |
| 1 | 1 | 1 | 1 | 1.44 | 0.15 | 1.37 | 0.60 | 0.64 | 0.11 | −24 | −9 | 0 | 9 | 47 | −23 | 5.04 |
| 2 | 1 | 1 | 1 | 3.40 | 0.16 | 1.80 | 0.92 | 0.68 | 0.03 | 76 | −1 | 41 | 52 | 79 | −247 | 5.59 |
| 2 | 1 | 1 | 2 | 3.32 | 0.14 | 1.50 | 0.87 | 0.70 | 0.15 | 11 | −27 | −6 | 14 | 31 | −22 | 6.61 |
| 2 | 2 | 1 | 2 | 2.32 | 0.24 | 0.90 | 0.66 | 0.80 | 0.18 | −11 | 1 | −31 | 2 | 47 | −8 | 9.19 |
| 2 | 2 | 2 | 2 | 2.22 | 0.20 | 3.29 | 0.43 | 0.26 | 0.20 | −7 | −8 | 28 | −7 | 0 | −5 | 15.31 |
| 2 | 3 | 2 | 2 | 2.04 | 0.26 | 3.34 | 0.48 | 0.28 | 0.15 | −11 | 2 | 31 | −4 | 2 | −19 | 16.25 |
| 2 | 3 | 2 | 3 | 2.76 | 0.27 | 3.34 | 0.59 | 0.28 | 0.18 | −3 | −1 | 22 | −1 | −2 | −15 | 17.06 |
| LSD (0.05) | | | | | | | | | | | | | | | | 0.60 |
| LSD (0.01) | | | | | | | | | | | | | | | | 0.82 |

[a]From Sumner (1981).

In reality, however, the diagnostic failures resulted from a lack of input data. In other words, in plot $N_2P_2K_2S_1$, when DRIS determined N to be most needed, some other factor was actually limiting yield. In Table 7, foliar analysis data from the same experiment were rediagnosed using tissue S, Mg, and Ca data as well as N, P, and K. In this set of diagnoses, it can be seen that S was actually more limiting than P in the $N_2P_1K_1S_1$ treatment. Whereas in the first set of diagnoses the yield peaked at 59% of the maximum attained in this experiment, expanding the diagnoses to include S, Ca, and Mg allowed the yield to be increased to 100%. This is corroborated by the data in Table 6, where addition of S as a basal treatment enabled a correct diagnosis to be made based only on N, P, K indices because S was no longer limiting. Therefore, it should be borne in mind that failure to make a correct diagnosis can result from consideration of too few factors.

## VIII. Comparisons of DRIS and Other Diagnostic Systems

It has been demonstrated with a number of crop species that DRIS can identify yield-limiting nutrient components and order them in terms of severity to allow corrective treatments to be applied. Comparisons of DRIS and other diagnostic systems, most notably critical value and sufficiency range systems, have also been made. A strict comparison of the accuracy of these various systems is difficult because only DRIS is presented as an integrated package. The critical value and sufficiency range systems are general approaches with no specific guidelines for standard value generation, although the "accuracy" of both of these systems is, to some degree, dependent upon this process. Wide ranges in standard values have been published for both the critical value and the sufficiency range systems; so, for the purpose of making comparative diagnoses, generally accepted or widely used standard values are selected. Although in theoretical terms this may be somewhat conservative (that is, these may not be the "best" available norms), in a real sense the use of these standards represents the limits of these systems as they are presently practiced. With this in mind, some previously published comparative diagnoses will be reviewed.

In most comparisons of diagnostic capabilities of critical value or sufficiency range systems and DRIS, tissue sampling has been carried out in such a manner as to satisfy the narrow range of conditions usually dictated by the critical value and sufficiency range systems (i.e., a specific stage of growth and leaf position). Even under these conditions, DRIS usually maintains slightly higher diagnostic precision. For example, Elwali and Gascho (1984) reported that sums of DRIS indices irrespective of sign for sugarcane were significantly decreased when fertilization was based on DRIS rather than on critical values (Table 8). Yields of both cane and sugar were also significantly improved when DRIS recommendations were followed.

**Table 8.** Effects of methods of guided fertilization on sum of DRIS indices irrespective of sign ($\Sigma I$) for three, five, and nine leaf nutrients in September and on yields of sugarcane[a]

| Methods of guided fertilization | $\Sigma I$ among: | | | Yield ($Mg\ ha^{-1}$) | |
|---|---|---|---|---|---|
| | NPK[b] | N–Mg[c] | N–Cu[d] | Cane | Sugar |
| Soil testing | 8 | 34 | 46 | 75.4 | 7.9 |
| Foliar analysis (CNL)[e] | 3 | 21 | 36 | 74.3 | 7.7 |
| Foliar analysis (DRIS) | 3 | 13 | 24 | 85.9 | 9.4 |
| $LSD_{0.05}$ | 6 | 11 | 10 | 5.3 | 0.7 |

[a]Elwali and Gascho (1984).

[b]N, P, and K indices only.

[c]N, P, K, Ca, and Mg indices.

[d]N, P, K, Ca, Mg, Fe, Mn, Zn, and Cu indices.

[e]CNL = critical nutrient level.

Similar results were obtained by Sumner (1979) for potatoes (Table 9). The DRIS-based treatments resulted in 39 successes or positive responses and 12 failures or negative responses, with a net yield response of +209.63 $Mg\ ha^{-1}$. Treatments based on critical values, on the other hand, resulted in 22 positive and 11 negative responses, for a net yield increase of 86.50 $Mg\ ha^{-1}$. The corresponding figures for sugarcane were 38 successes and 13 failures with DRIS, 20 successes and 9 failures when using critical values, with net yield responses of 305.42 $Mg\ ha^{-1}$ and 199.05 $Mg\ ha^{-1}$ for the two systems. For corn, 166 successes and 24 failures were recorded with DRIS, 133 successes and 34 failures with the critical value system. Net responses were 295.56 $Mg\ ha^{-1}$ and 220.81 $Mg\ ha^{-1}$, respectively, for the two systems with this crop.

Many other workers (listed below) have reported the superiority of DRIS over critical value, sufficiency range and other approaches:

| Source | Crop |
|---|---|
| Elwali and Gascho (1984) | Sugarcane |
| Escano, et al. (1981) | Corn |
| C. A. Jones and Bowen (1981) | Sugarcane |
| Langenegger and Smith (1978) | Pineapple |
| Meldel-Johnsen and Sumner (1980) | Potato |
| Meyer (1981) | Sugarcane |
| Sumner (1983) | Sugarcane |

The diagnosis and recommendation integrated system has the capability of making diagnoses in some cases when other systems cannot. In Table 10, for example, progressive diagnoses of potatoes from an NPK factorial experiment are based on DRIS and are contrasted with diagnoses based

**Table 9.** Comparison of the success and failures achieved by the critical value and DRIS approaches to foliar diagnosis of various crops grown in field experiments[a]

| | | DRIS | | | | Critical value | | | |
|---|---|---|---|---|---|---|---|---|---|
| | | Success[b] | | Failures[b] | | Successes | | Failures | |
| Crop | Nutrient applied | No. | Yield response ($Mg\ ha^{-1}$) | No. | Yield response ($Mg\ ha^{-1}$) | No. | Yield response ($Mg\ ha^{-1}$) | No. | Yield response ($Mg\ ha^{-1}$) |
| Potatoes[c] | N | 5 | 17.19 | 2 | −5.56 | 5 | 19.04 | 4 | −19.76 |
| | P | 19 | 81.61 | 9 | −33.14 | 13 | 54.47 | 7 | −18.89 |
| | K | 15 | 152.40 | 1 | −2.87 | 4 | 50.64 | 0 | 0.00 |
| | Total | 39 | 251.20 | 12 | −41.57 | 22 | 125.15 | 11 | −38.65 |
| Sugarcane[d] | N | 12 | 210.25 | 3 | −56.84 | 7 | 154.02 | 3 | −56.62 |
| | P | 6 | 95.10 | 1 | −5.58 | 0 | 0.00 | 0 | 0.00 |
| | K | 20 | 153.16 | 10 | −90.67 | 13 | 187.64 | 6 | −85.99 |
| | Total | 38 | 458.51 | 13 | −153.09 | 20 | 341.66 | 9 | −142.61 |
| Corn[e] | N | 47 | 61.70 | 8 | −3.41 | 31 | 32.34 | 9 | −4.62 |
| | P | 40 | 94.83 | 6 | −2.63 | 35 | 69.13 | 8 | −2.19 |
| | K | 79 | 151.06 | 10 | −5.99 | 67 | 136.59 | 17 | −10.44 |
| | Total | 166 | 307.59 | 24 | −12.03 | 133 | 238.06 | 34 | −17.25 |

[a]From Sumner (1979).

[b]A success is defined as the situation in which the addition of a nutrient diagnosed as limiting yield results in a yield increase. A failure is the reverse situation.

[c]Critical values from Lintner (1967).

[d]Critical values from Gosnell and Long (1971).

[e]Critical values from Melsted *et al.* (1969).

**Table 10.** Comparison of three methods of foliar diagnosis of nutrient requirements of potatoes using data from a $3^3$ NPK fertilizer experiment published by Lintner (1967), from Sumner (1979)

| Treatment | | | Leaf composition % | | | DRIS indices | | | Tuber Yield | Method of diagnosis[a] | | |
|---|---|---|---|---|---|---|---|---|---|---|---|---|
| | | | | | | | | | | Order of requirement | Normal range Geraldson | Critical value |
| N | P | K | N | P | K | N | P | K | ($Mg\ ha^{-1}$) | by DRIS | *et al.* (1973) | Lintner (1967) |
| 2 | 0 | 0 | 5.35 | 0.23 | 2.47 | 42 | −13 | −29 | 10.9 | K > P > N | K | P, K |
| 2 | 0 | 1 | 5.47 | 0.25 | 5.67 | 18 | −23 | 5 | 22.2 | P > K > N | nd | nd |
| 2 | 1 | 1 | 5.12 | 0.25 | 4.76 | 18 | −13 | 0 | 27.9 | P > K > N | nd | nd |
| 2 | 1 | 2 | 5.35 | 0.25 | 9.36 | 8 | −35 | 27 | 24.4 | P > N > K | nd | nd |
| 2 | 2 | 1 | 5.13 | 0.26 | 3.60 | 24 | −12 | −12 | 28.1 | K = P > N | nd | nd |
| 2 | 2 | 2 | 5.20 | 0.25 | 5.89 | 15 | −22 | 7 | 32.0 | P > K > N | nd | nd |
| $LSD_{0.05}$ | | | 0.46 | 0.02 | 1.97 | | | | 3.5 | | | |

[a]nd = no diagnosis possible.

**Table 11.** Comparison of three methods of foliar diagnosis of nutrient requirements of corn using data from two experiments in Illinois supplied by T. R. Peck (personal communication), from Sumner (1979)

| Treatment[a] | | | | Leaf composition (%) | | | | | DRIS indices | | | | | Grain yield | Method of diagnosis[b] | | |
|---|---|---|---|---|---|---|---|---|---|---|---|---|---|---|---|---|---|
| N | P | K | L | N | P | K | Ca | Mg | N | P | K | Ca | Mg | Mg ha$^{-1}$ | Order of requirement DRIS | Sufficiency range, J.B. Jones and Eck (1973) | Critical value, Melsted *et al.* (1969) |
| Aledo experiment 1968 (100 series) | | | | | | | | | | | | | | | | | |
| 2 | 0 | 2 | 0 | 3.66 | 0.26 | 2.17 | 0.57 | 0.26 | 12 | −10 | 3 | 0 | −5 | 5.303 | P > Mg > Ca > K > N | nd | nd |
| 2 | 1 | 2 | 0 | 3.82 | 0.35 | 2.08 | 0.79 | 0.30 | 5 | 0 | −5 | 06 | −6 | 8.821 | Mg > K > P > N > Ca | nd | nd |
| 2 | 1 | 2 | 1 | 3.76 | 0.32 | 1.82 | 0.76 | 0.32 | 8 | −2 | −8 | 05 | −3 | 9.151 | K > Mg > P > Ca > N | nd | K |
| Aledo experiment 1968 (200 series) | | | | | | | | | | | | | | | | | |
| 1 | 0 | 1 | 1 | 3.37 | 0.26 | 1.96 | 0.85 | 0.33 | 5 | −11 | −3 | 10 | −1 | 7.617 | P > K > Mg > N > Ca | nd | nd |
| 1 | 1 | 1 | 1 | 3.63 | 0.37 | 2.52 | 0.86 | 0.20 | 3 | 3 | 3 | 13 | −22 | 9.910 | Mg > N = P = K > Ca | nd | nd |

[a] $N_1$ = 87 kg N ha$^{-1}$, $N_2$ = 175 kg N ha$^{-1}$, $P_0$ = 0, $P_1$ = 33 kg P ha$^{-1}$, $P_3$ = 66 kg P ha$^{-1}$, $K_1$ = 44 kg K ha$^{-1}$, $K_1$ = 88 kg K ha$^{-1}$, $L_0$ = 0, $L_1$ = lime to pH 6.2.

[b] nd = no diagnosis possible.

on normal nutrient ranges and critical values. At the lowest treatment level ($N_2P_0K_0$), the tissue K was below the normal range and the critical value, and K was determined to be most needed by all three systems. However, diagnoses of plants from subsequent treatments were meaningful only with DRIS. Nitrogen, P, and K were above the critical values as well as the lower limits of the normal ranges in these cases, but yields continued to increase as these nutrients were added in accordance with DRIS. An example using sufficiency ranges and critical values to diagnose corn ear leaf tissue is illustrated in Table 11. Again, DRIS diagnoses led to substantial yield increases through improved nutrient balance even when all nutrient levels were above critical values and within nutrient sufficiency ranges.

# IX. Effect of Leaf Age and Position on DRIS Indices

Although DRIS-based recommendations are generally at least as accurate those based on other systems when sampling is conducted under the conditions prescribed by those other systems, a major strength of DRIS is its ability to diagnose plants sampled at various growth stages. It was earlier demonstrated that nutrient ratios were frequently less affected by plant age than nutrient concentrations based on dry matter (Figure 2 and Table 2). The DRIS indices based on properly selected nutrient expressions should, therefore, also show reduced dependence on plant age. This is, in fact, the case and is illustrated by the data in Table 12, where the order of relative abundance of N, P, and K in sugarcane tissue remained constant over a 10-month sampling period. Furthermore, the inflexibility of the critical value system is illustrated by the fact that no insufficiencies were indicated in the sugarcane tissue until the fifth month, and the nutrient deemed most needed by DRIS (P) was not diagnosed as insufficient until the seventh month. The N, P, and K DRIS indices in wheat (*Triticum aestivum* L.) and corn tissue also exhibit limited sensitivity to plant age (Sumner, 1977b, c).

When such nutrients as Ca and Mg, the concentrations of which tend to increase with advancing maturity, are related to such nutrients as N, P, and K, which generally decrease with time, ratios may not be the appropriate form of expression (see Figure 2). Products formed between nutrients from these two groups are likely to be the most constant form of expression for these situations (note the stability of N $\times$ Ca in Figure 2).

When nutrient products are used instead of nutrient ratios, some adjustments are necessary in the calculation of nutrient indices. If, for example, N $\times$ Ca were used, a new nutrient $(X)$ having the value 1/Ca would have to be defined. Thus, N $\times$ Ca = N/(1/Ca) = N/$X$. One then proceeds through the index calculations using N/$X$ as the ratio for both the unknown

**Table 12.** Effect of age on foliar diagnosis of the nutrient requirements of sugarcane using data published by Gosnell and Long (1971), from Sumner (1979)

| Crop age | Leaf composition (%) | | | DRIS indices | | | Method of diagnosis | |
|---|---|---|---|---|---|---|---|---|
| | | | | | | | Order of | Critical value, |
| months | N | P | K | N | P | K | requirement by DRIS | Gosnell and Long (1971)[a] |
| 1 | 2.70 | 0.27 | 1.65 | 9 | −9 | 0 | P > K > N | nd |
| 2 | 2.22 | 0.23 | 1.52 | 5 | −10 | 5 | P > N = K | nd |
| 3 | 1.99 | 0.21 | 1.48 | 2 | −11 | 9 | P > N > K | nd |
| 4 | 1.86 | 0.20 | 1.50 | −1 | −12 | 13 | P > N > K | nd |
| 5 | 1.78 | 0.18 | 1.48 | 0 | −17 | 17 | P > N > K | N |
| 6 | 1.68 | 0.18 | 1.42 | −2 | −14 | 16 | P > N > K | N |
| 7 | 1.68 | 0.17 | 1.46 | −1 | −18 | 19 | P > N > K | N, P |
| 8 | 1.68 | 0.17 | 1.44 | −1 | −18 | 19 | P > N > K | N, P |
| 9 | 1.62 | 0.17 | 1.50 | −4 | −18 | 22 | P > N > K | N, P |
| 10 | 1.56 | 0.16 | 1.40 | −2 | −19 | 21 | P > N > K | N, P |

[a] nd = no diagnosis possible.

**Table 13.** Effect of leaf number and leaf part sampled on the concentration of N, P, K, Ca, and Mg and calculated DRIS indices for corn[a]

| Leaf number below tassel | Leaf composition (%) | | | | | DRIS indices | | | | | Order of requirement |
|---|---|---|---|---|---|---|---|---|---|---|---|
| | N | P | K | Ca | Mg | N | P | K | Ca | Mg | |
| 1 | 1.60 | 0.25 | 1.35 | 0.83 | 0.56 | −35 | 1 | −11 | 15 | 29 | N > K > P > Ca > Mg |
| 3 | 2.25 | 0.27 | 1.37 | 0.78 | 0.57 | −15 | −2 | −15 | 8 | 24 | N = K > P > Ca > Mg |
| 6 | 2.35 | 0.33 | 1.77 | 0.86 | 0.67 | −23 | 2 | −10 | 6 | 25 | N > K > P > Ca > Mg |
| 7 | 2.35 | 0.29 | 1.88 | 0.73 | 0.52 | −16 | −2 | −3 | 4 | 16 | N > K > P > Ca > Mg |
| 9 | 2.30 | 0.25 | 2.02 | 0.81 | 0.57 | −17 | −11 | 1 | 7 | 21 | N > P > K > Ca > Mg |

[a]Data from J.B. Jones, 1970; from Sumner 1977c

sample and the norm. In this case reverse sign notation for the functions in the index equations would be used.

The following is presented as a practical illustration of the usefulness of product forms of nutrient expression. Beverly *et al.* (1984) diagnosed N, P, K, Ca, and Mg status in the leaves of Valencia oranges (*Citrus sinensis* L.) using DRIS, wherein these five nutrients were all expressed in ratio form, and reported that nutrient indices were affected by leaf age. In contrast, Sumner (1985) analyzed the same data expressing Ca and Mg as products with N, P, and K (N × Ca rather than N/Ca or Ca/N, etc.) and found that leaf age had no substantial effect on DRIS nutrient diagnoses of leaves from orange trees. Therefore, proper inputs must be provided if one is to take full advantage of the flexibility that DRIS can provide.

The position of sampled leaves on plants may also have a limited impact on diagnostic results when DRIS is used. When diagnoses of corn leaves varying in their position on the plant were carried out with DRIS, only minor variations in nutrient order occurred and the nutrient diagnosed as most needed was largely independent of leaf position (Table 13). An example of DRIS analysis of soybean (*Glycine max* L. Merrill) foliage, representing plants ranging widely in age and from various locations on the plant, is shown in Table 14 (Sumner, 1977e). Again, very little change in order of nutrients occurred, even though the range of plant age was ap-

**Table 14.** Effect of stage of growth and position of leaf sampled on leaf composition and DRIS indices for soybeans[a]

| Stage of growth | Days after emergence (approx.) | Leaf composition (%) | | | DRIS indices | | | Order of nutrient requirement |
|---|---|---|---|---|---|---|---|---|
| | | N | P | K | N | P | K | |
| Lower leaves (seven lowest nodes) | | | | | | | | |
| 1 | 20 | 5.10 | 0.23 | 1.40 | 27 | −15 | −12 | P > K > N |
| 2 | 26 | 5.20 | 0.35 | 1.65 | 9 | 3 | −12 | K > P > N |
| 3 | 40 | 5.30 | 0.31 | 1.50 | 17 | −1 | −16 | K > P > N |
| 5 | 52 | 4.80 | 0.24 | 1.00 | 35 | −3 | −32 | K > P > N |
| 7 | 73 | 4.00 | 0.24 | 1.00 | 21 | 3 | −24 | K > P > N |
| 9 | 92 | 3.10 | 0.20 | 1.05 | 9 | −1 | −8 | K > P > N |
| 10 | 102 | 2.20 | 0.16 | 0.75 | 5 | 5 | −10 | K > P = N |
| Middle leaves (8 to 14 nodes) | | | | | | | | |
| 5 | 52 | 5.90 | 0.39 | 1.50 | 18 | 8 | −26 | K > P > N |
| 7 | 73 | 5.30 | 0.31 | 1.40 | 20 | 0 | −20 | K > P > N |
| 9 | 92 | 4.00 | 0.23 | 1.30 | 13 | −5 | −8 | K > P > N |
| 10 | 102 | 2.95 | 0.22 | 1.15 | 1 | 3 | −4 | K > N > P |
| Top leaves (above node 14) | | | | | | | | |
| 7 | 73 | 6.45 | 0.42 | 1.75 | 15 | 6 | −21 | K > P > N |
| 9 | 92 | 4.25 | 0.25 | 1.10 | 20 | 1 | −21 | K > P > N |
| 10 | 102 | 2.90 | 0.20 | 0.95 | 8 | 3 | −11 | K > P > N |

[a]Data interpolated from Figure 4, Hanway and Weber (1971), from Sumner (1977d).

proximately 2 months. Therefore, in many cases DRIS can minimize sampling constraints, which are generally considered to be the most severe limitations of other diagnostic systems.

## X. Universality of Foliar Norms

If foliar norms developed under one set of conditions are to be applied to another, the effect of changing conditions on nutritional optima must be known. In other words, the elemental composition of extremely high-yielding plants must be nearly identical regardless of geographical or climatic origin if norms are to be applied universally. Unfortunately, very little effort has been devoted to studies of such relationships.

When DRIS norms developed from various areas are compared, regional differences sometimes become apparent. In Table 15 sugarcane norms derived from plants grown on organic Florida soils are contrasted with those developed from plants grown on mineral South African soils. Despite the tremendous differences in soil and environmental conditions the nutrient norms derived by averaging values from these two groups of high-yielding plants remained virtually unchanged.

Comparison of corn ear leaf norms developed from various parts of the world, however, reveal some substantial differences, particularly where Mg and Ca nutrition are concerned (Table 16). Corn tissue from the Southeastern United States, South Africa, and, to a lesser extent, from Hawaii (areas with highly weathered, low cation-exchange capacity soils) contained lower levels of Ca than did foliage from plants grown in other areas. Furthermore, Mg levels were somewhat low in the sample base from the Southeastern United States. Calcium and Mg norms for soybeans and alfalfa seem to be similarly affected by soil and/or climatic conditions in comparisons between plants grown in the Midwestern and Southeastern sections of the United States (Tables 17 and 18).

**Table 15.** Sugarcane foliar norms from Florida and South Africa

| | Florida (Elwali and Gascho, 1983) | South Africa (Beaufils and Sumner, 1976) |
|---|---|---|
| N/P | 8.706 | 8.197 |
| N/K | 1.526 | 1.511 |
| K/P | 5.633 | 5.464 |
| Ca/N | 0.151 | 0.128 |
| Ca/P | 1.314 | 1.146 |
| Ca/K | 0.222 | 0.205 |
| Ca/Mg | 1.373 | 1.158 |
| Mg/N | 0.113 | 0.116 |
| Mg/P | 0.984 | 0.962 |
| Mg/K | 0.163 | 0.186 |

**Table 16.** Maize ear leaf tissue norms from various sources

| | Combined data base | South Africa | Southeastern United States[a] | Northeastern United States and Canada[b] | Midwestern United States[c] | New Zealand[d] | Hawaii[e] |
|---|---|---|---|---|---|---|---|
| N (%) | 3.26 | 3.16 | 3.34 | 3.04 | 3.29 | — | — |
| N/P | 10.13 | 8.91 | 11.23 | 9.14 | 9.96 | 11.54 | 9.98 |
| N/K | 1.40 | 1.32 | 1.24 | 1.43 | 1.50 | 1.30 | 1.60 |
| N/Ca | 6.95 | 7.96 | 8.31 | 5.92 | 5.80 | 5.13 | 7.04 |
| N/Mg | 16.97 | 17.28 | 20.10 | 13.03 | 15.13 | 15.62 | 13.51 |
| N/S | 12.17 | 11.04 | 15.02 | 11.45 | 12.86 | — | 15.20 |
| P (%) | 0.330 | 0.369 | 0.303 | 0.339 | 0.338 | — | — |
| P/K | 0.146 | 0.155 | 0.115 | 0.164 | 0.154 | 0.108 | 0.163 |
| P/Ca | 0.678 | 0.922 | 0.748 | 0.640 | 0.570 | 0.451 | 0.725 |
| P/Mg | 1.65 | 2.00 | 1.80 | 1.42 | 1.53 | 1.41 | 1.37 |
| P/S | 1.33 | 1.28 | 1.61 | 1.27 | 1.30 | — | 1.55 |
| K (%) | 2.42 | 2.46 | 2.78 | 2.19 | 2.24 | — | — |
| K/Ca | 5.32 | 6.33 | 6.88 | 4.35 | 3.91 | 3.97 | 4.46 |
| K/Mg | 11.95 | 13.82 | 16.65 | 9.68 | 10.45 | 11.90 | 8.55 |
| K/S | 8.74 | 8.70 | 12.04 | 8.26 | 8.73 | — | 9.67 |
| Ca (%) | 0.531 | 0.442 | 0.431 | 0.595 | 0.602 | — | — |
| Ca/Mg | 2.43 | 2.07 | 2.49 | 2.15 | 2.59 | 2.95 | 1.96 |
| Ca/S | 2.18 | 1.55 | 2.07 | 2.24 | 2.17 | — | 2.21 |
| Mg (%) | 0.241 | 0.245 | 0.176 | 0.312 | 0.258 | — | — |
| Mg/S | 1.02 | 0.843 | 1.06 | 1.17 | 0.846 | — | 1.18 |
| S (%) | 0.274 | 0.289 | 0.247 | 0.274 | 0.269 | — | — |

[a]Includes data from Alabama, Georgia, North Carolina, and Virginia.

[b]Includes data from Delaware, Maryland, New Jersey, Pennsylvania, and Ontario, Canada.

[c]Includes data from Illinois, Indiana, Iowa, Michigan, Minnesota, Ohio, and Wisconsin.

[d]From Cornforth and Steele (1981).

[e]From Escano *et al.* (1981).

**Table 17.** Soybean foliar norms from the Midwestern and Southeastern United States[a]

| | Midwest | Southeast |
|---|---|---|
| N/P | 14.9 | 15.3 |
| N/K | 2.69 | 2.60 |
| N/Ca | 3.98 | 5.44 |
| Mg/N | 0.0871 | 0.0604 |
| P/K | 0.176 | 0.170 |
| P/Ca | 0.291 | 0.357 |
| P/Mg | 0.915 | 1.12 |
| K/Ca | 1.74 | 2.18 |
| Mg/K | 0.209 | 9.154 |
| Mg/Ca | 0.315 | 0.326 |

[a]From Beverly *et al.* (1985).

These phenomena may place some restraints on the universal application of norms developed in one locale, at least with regard to Mg and Ca, depending on the breadth of available data bases. Presumably, the variations illustrated above reflect the wide range of permissable Ca and Mg levels in healthy or high-yielding plants, which is corroborated by the relative values of the CVs associated with ratios of various nutrients from a high-yielding population. For example, the CV that corresponds to the Mg/DM norm from a worldwide corn data base (containing over 8000 ob-

**Table 18.** Alfalfa foliar norms from the Midwestern and the Southeastern United States

| | Southeast (Walworth *et al.*, 1986b) | Midwest (Erickson *et al.*, 1982) |
|---|---|---|
| N (%) | 2.95 | 3.29 |
| N/P | 12.45 | 10.30 |
| N/K | 1.50 | 1.26 |
| N/Ca | 2.53 | 2.46 |
| Mg/N | 0.055 | 0.086 |
| P (%) | 0.244 | 0.320 |
| P/K | 0.124 | 0.124 |
| P/CA | 0.216 | 0.230 |
| Mg/P | 0.672 | 0.924 |
| K (%) | 2.03 | 2.67 |
| K/Ca | 1.94 | 1.87 |
| Mg/K | 0.083 | 0.113 |
| Ca (%) | 1.19 | 1.36 |
| Mg/Ca | 0.137 | 0.113 |
| Mg (%) | 0.161 | 0.290 |

servations) is 42%, whereas those associated with N/DM, P/DM, and K/DM are 12, 30, and 32%, respectively.

If the data bases used to develop norms are broad enough to encompass the permissable variation, no serious problem should exist. The variation is taken into account in the index equations where the deviation of each nutrient expression from its optimum value is weighted by the inverse of the corresponding CV. If all expressions involving a given nutrient have high CVs, therefore, the nutrient index tends toward zero. Therefore, numerical expressions of nutrients exhibiting a wide range of values consistent with high yields must deviate substantially from their optima to influence the resultant nutrient indices.

The main danger, then, in diagnosing plants from one geographical region with norms and CVs developed in another is that the data may be skewed, and the CVs may not reflect the extent of normal variation. If data are pooled from various areas, however, such that normal variation is represented, those values should then be applicable to a range of specific locales and conditions.

## XI. Expansion of DRIS Beyond Nutrient Ratios

The diagnosis and recommendation integrated system can be expanded to evaluate data other than those derived from foliar analyses. For instance, Beaufils and Sumner (1976) developed DRIS norms for soil test P, K, Ca, and Mg to be applied to sugarcane culture on South African soils. As with foliar nutrient values, soil test levels of the various nutrients are ratioed, and norms generated by averaging values from those observations associated with high yields. Coefficients of variation are also generated from these data and DRIS indices are calculated in a manner identical to that described for plant tissue data.

As with foliar diagnoses, the use of DRIS with soil data provides the advantage of taking into account nutrient balance and ranking nutrients in terms of abundance relative to optimal levels. The concept of an ideal or optimum soil nutrient balance has been promoted previously under what is sometimes known as the "Basic Cation Saturation Ratio" system (McLean, 1977). This concept, originated by Bear and co-workers (Bear *et al.*, 1944, 1945; Bear and Prince, 1945; Bear and Toth, 1948), advocates the use of specific fractional levels of nutrient saturation of cation-exchange capacity (CEC), i.e., 65% Ca, 10% Mg, etc., rather than nutrient ratios. Although this seems a subtle distinction, the difference may take on importance if the measured CEC is not that existing under field conditions, as is often the case when extractants buffered at a high pH are used in CEC determination on soils with significant variable-charge components.

Recently, the balance or ratio concept for optimization of soil nutrients has received criticism because a wide range of saturation ratios seems to support maximum yields in field trials (Liebhardt, 1981; McLean *et al.*, 1983). However, yield levels in such field experiments have often been quite low. McLean *et al.* (1983) recorded maximum yields of 6.3 and 2.7 Mg $ha^{-1}$ for corn and wheat, respectively, in experiments used to illustrate the wide range of nutrient saturation ratios consistent with "high" yields. Examination of the yield response model represented in Figures 3 and 4 reveals that the permissable range of any independent variable consistent with a given yield level is greater when yield is low. Certainly, a range exists even at very high yield levels, but one should not be surprised at the breadth of permissable ranges consistent with low yields. The use of DRIS or other nutrient balance systems to analyze soil data needs further investigation, preferably in high-yield situations where the benefits of such schemes are likely to be most apparent.

The diagnosis and recommendation integrated system may also be expanded to include expressions of nonessential elements, such as Si or Na, or nonnutritional variables, such as plant population or planting date, although such constituents have not been included in published calibrations or diagnoses. Theoretically, such nutrient forms as nitrate and ammonium could be considered separately and treated as individual nutritional entities within DRIS expressions. Again, no diagnoses of this type have been published.

However, DRIS has recently been expanded to include nutrient ratios with dry matter (DM) (Walworth *et al.*, 1986b). These expressions are, of course, identical to what are commonly termed nutrient concentrations (macronutrients are usually expressed as percentages, whereas micronutrients are expressed as parts per million of dry plant material). In this case, dry matter is treated as an additional plant constituent and an index is calculated for dry matter as for other plant constituents. In fact, dry matter is essentially the sum of three nutrients that are usually ignored in nutritional considerations, namely C, H, and O. In a dried corn leaf, for instance, approximately 6 to 8% of the total mass is attributable to inorganic constituents, with the remaining 92 to 94% being composed of C, H, and O. Therefore, the dry matter index of a plant should be representative of the processes of C, H, and O aquisition. Although seldom discussed in such terms, an individual nutrient concentration, such as N% (100N/DM), is a measure of N acquisition relative to the accumulation of C, H, and O and is no more reflective of the former process than of the latter.

To include dry matter in DRIS calculations, expressions relating each nutrient to dry matter are included in the nutrient index equations. For example, if indices were to be calculated for N, P, K, and DM for a given sample the ratios used might be N/P, N/K, N/DM, P/K, P/DM, and K/DM, where N/DM, P/DM, and K/DM are simply N% ÷ 100, P% ÷ 100, and K% ÷ 100, respectively. Of course, if these values have already been calculated as percentages, there is no need to change them, for division by the norms in the function equations results in unitless values. The N index, as before, would be calculated as $[f(N/P) + f(N/K) + f(N/DM)]/3$ and P and K indices are calculated similarly. The DM index would be equal to $[-f(N/DM) - f(P/DM) - f(K/DM)]/3$ in like fashion. A ranking is thus obtained describing the relative abundance of N, P, K, and DM, and the task remains of interpreting this ranking.

Using the critical value system, N, for instance, would be deemed lacking if it were in relatively shorter supply than DM (if N% or N/DM were less than the critical value). Analogously, N might be considered lacking if its DRIS index were less than that of DM. Analysis of corn yield responses in a factorial field trial was based on this type of diagnosis (Sumner, 1963, unpublished data). When nutrients with indices less than that of DM were added, yield increases were noted 74% of the time, whereas yield declines occurred in only 26% of the cases tested. Conversely, if nutrients with indices greater than that of DM were added, yield increased in 48% of the cases and declined in the remaining 52%. Therefore, the DM index may have a useful application as a delimiter between nutrients that are present in inadequate quantities for the amount of C, H, and O acquisition occuring in a plant, and those which are adequate for the existing rate of that process. However, the practical application of this theory remains to be tested in an extensive range of field situations.

## XII. Conclusions

The strengths of DRIS, including reduced effect of tissue age and placement, have been demonstrated. There is considerable room for expansion of DRIS, with respect to both broadened data for various crop plants and previously neglected growth factors. To date, DRIS norms have been published for a wide range of plants, and as a convenience to the reader citations for the various crops are presented in Table 19. Norms for some of these species are based on somewhat limited data. Expansion of these data bases and further validation of existing norms is a primary prerequisite for widespread routine use of DRIS. There is particularly a need for data representing plants at various stages of growth, and for evaluation of these data with respect to stability of various forms of nutrient expressions that will ultimately allow greater flexibility in tissue sampling and wider applicability of subsequent recommendations.

The diagnosis and recommendation integrated system represents a step forward in our abilities to diagnose nutritional plant conditions and may ultimately have a significant impact on agricultural practices. However, as are other diagnostic systems, DRIS is dependent on the quality of the empirically determined information that is its input.

**Table 19.** Citations for published DRIS norms

| Crop | Source |
| --- | --- |
| Alfalfa (*Medicago sativa* L.) | Kelling *et al.* (1983) |
| Alfalfa (*Medicago sativa* L.) | Walworth *et al.* (1986b) |
| Corn (*Zea mays* L.) | Sumner (1981) |
| Corn (*Zea mays* L.) | Escano *et al.* (1981) |
| Corn (*Zea mays* L.) | Elwali *et al.* (1985) |
| Citrus (*Citrus sinensis* L.) | Beverley *et al.* (1984) |
| Citrus (*Citrus sinensis* L.) | Sumner (1985) |
| Oats (*Avena sativa*) | Chojnacki (1984) |
| Peaches (*Prunus persica* Batsch) | Sumner (1985) |
| Pineapples (*Ananas comosus* L. Merr.) | Langenegger and Smith (1978) |
| Poplars (*Populus* spp.) | Leech and Kim (1981) |
| Potatoes (*Solanum tuberosum*) | Meldal-Johnsen and Sumner (1980) |
| Rubber (*Hevia brasiliensis* H. B. K.) | Beaufils (1957) |
| Soybeans (*Glycine max*. L. Merrill) | Sumner (1977e) |
| Soybeans (*Glycine max*. L. Merrill) | Hallmark *et al.* (1984, 1985) |
| Sugarcane (*Saccarum officinarum* L.) | Beaufils and Sumner (1976) |
| Sugarcane (*Saccarum officinarum* L.) | Elwali and Gascho (1983, 1984) |
| Sunflower (*Helianthus annus* L.) | Grove and Sumner (1982) |
| Tea (*Camelia senensis* L. Ktze) | Lee (1980) |
| Wheat (*Triticum aestivum* L.) | Sumner (1981) |

## References

Aldrich, S.R. 1973. Plant analysis: Problems and Opportunities. In: L.M. Walsh, and J.D. Beaton (eds.) *Soil Testing and Plant Analysis*. Soil Science Society of America, Madison, WI, pp. 213–221.

Bates, T.E. 1971. Factors affecting critical nutrient concentrations in plants and their evaluation: A review. *Soil Sci.* 112:116–130.

Bear, F.E., and A.L. Prince. 1945. Cation-equivalent constancy in alfalfa. *J. Am. Soc. Agron.* 37:217–222.

Bear, F.E., and S.J. Toth. 1948. Influence of calcium on the availability of other soil cations. *Soil Sci.* 65:69–74.

Bear, F.E., A.L. Prince, and J.L. Malcolm. 1944. The potassium-supplying powers of 20 New Jersey soils. *Soil Sci.* 58:139–149.

Bear, F.E., A.L. Prince, and J.L. Malcolm. 1945. Potassium needs of New Jersey soils. NJ Agricultural Experiment Station Bull. 721, Rutgers University, New Brunswick, NJ.

Beaufils, E.R. 1957. Research for rational exploitation of Hevea using a physiological diagnosis based on the mineral analysis of various parts of the plants. *Fertilite* 3:27–38.

Beaufils, E.R. 1971. Physiological diagnosis—A guide for improving maize production based on principles developed for rubber trees. *Fertil. Soc. S. Afr. J.* 1:1–28.

Beaufils, E.R. 1973. Diagnosis and recommendation integrated system (DRIS). Soil Science Bull. No. 1, University of Natal, S. Africa.

Beaufils, E.R., and M.E. Sumner. 1976. Application of the DRIS approach for calibrating soil, plant yield and plant quality factors of sugarcane. *Proc. S. Afr. Sugar Tech. Assoc.* 50:118–124.

Beverly, R.B., J.C. Stark, J.C. Ojala, and T.W. Embleton. 1984. Nutrient diagnosis of 'Valencia' oranges by DRIS. *J. Am. Soc. Hort. Sci.* 109:649–654.

Beverly, R.B., M.E. Sumner, W.S. Letzsch, and C.O. Plank. 1985. Foliar diagnosis of soybeans by DRIS. *Commun. Soil Sci. Plant Anal.* 17:3.

Bouma, D. 1983. Diagnosis of mineral deficiencies using plant tests. In: A. Lauchli and Bieleski, R.L. (eds.) *Encyclopedia of Plant Physiology.* Vol. 15A. *Inorganic Plant Nutrition.* Springer-Verlag, Berlin, pp. 120–146.

Chojnacki, A. 1984. The evaluation of the nutritional status of oats by the DRIS method. *Proc. VI Int. Colloq. Opt. Plant Nutr.* 1:139–148.

Cornforth, I.S., and K.W. Steele. 1981. Interpretation of maize leaf analysis in New Zealand. *N. Z. J. Expt. Agric.* 9:91–96.

Dow, A.I., and S. Roberts. 1982. Proposal: Critical nutrient ranges for crop diagnosis. *Agron. J.* 74:401–403.

Elwali, A.M.O., and G.J. Gascho. 1983. Sugarcane response to P, K and DRIS corrective treatments on Florida Histosols. *Agron. J.* 75:79–83.

Elwali, A.M.O., and G.J. Gascho. 1984. Soil testing, foliar analysis, and DRIS as guides for sugarcane fertilization. *Agron. J.* 76:466–470.

Elwali, A.M.O., G.J. Gascho, and M.E. Sumner. 1985. Sufficiency levels and DRIS norms for 11 nutrients in corn. *Agron. J.* 77:506–508.

Erickson, T., K.A. Kelling, and E.E. Schulte. 1982. Predicting alfalfa nutrient needs through DRIS. *Proc. Wisconsin Fertil. Aglime. & Pest. Management Conference.* 21:233–246.

Escano, C.R., C.A. Jones, and G. Uehara. 1981. Nutrient diagnosis in corn grown on Hydric Dystrandepts: II. Comparison of two systems of tissue diagnosis. *Soil Sci. Soc. Am. J.* 45:1140–1144.

Geraldson, C.M., G.R. Klacan, and O.A. Lorenz. 1973. Plant analysis as an aid in fertilizing vegetable crops. In: L.M.Walsh, and J.D. Beaton (eds.) *Soil Testing and Plant Analysis.* Soil Science Society of America, Madison, WI, pp. 365–379.

Gosnell, J.M., and A.C. Long. 1971. Some factors affecting foliar analysis in sugarcane. *Proc. S. Afr. Sugar Tech. Assoc.* 45:217–232.

Grove, J.H., and M.E. Sumner. 1982. Yield and leaf composition of sunflower in relation to N, P, K and lime treatments. *Fertil. Res.* 3:367–378.

Hallmark, W.B., J.F. Adams, and H.F. Morris. 1984. Detection of zinc deficiency in soybeans by the Diagnosis and Recommendation Integrated System. *J. Fertil. Iss.* 1:104–109.

Hallmark, W.B., J.F. Adams, and H.F. Morris. 1985. The use of plant analysis to diagnose nutrients limiting soybean yield for the subsequent year. *J. Fertil. Iss.* 2:66–73.

Hanway, J.J., and C.R. Weber. 1971. N, P and K percentages in soybean (*Glycine max* L. Merrill) plant parts. *Agron. J.* 63:286–290.

Jones, C.A. 1981. Proposed modifications of the Diagnosis and Recommendation Integrated System (DRIS) for interpreting plant analyses. *Commun. Soil Sci. Plant Anal.* 12:785–794.

Jones, C.A., and J.E. Bowen. 1981. Comparative DRIS and crop log diagnosis of sugarcane tissue analysis. *Agron. J.* 73:941–944.

Jones, J.B., Jr. 1970. Distribution of 15 elements in corn leaves. *Commun. Soil Sci. Plant Anal.* 127–134.

Jones, J.B., Jr., and H.V. Eck. 1973. Plant analysis as an aid in fertilizing corn and grain sorghum. In: L.M. Walsh, and J.D. Beaton (eds.) *Soil Testing and Plant Analysis*. Soil Science Society of America, Madison, WI, pp. 349–364.

Kelling, K.A., E.E. Schulte, and T. Kaehler. 1983. Use of DRIS in routine plant analysis. *Proc. Wisc. Fertil. Aglime Pest Management Conference* 22:239–243.

Langenneger, W., and B.L. Smith. 1978. An evaluation of the DRIS system as applied to pineapple leaf analysis. *Proc. 8th Int. Colloq. Plant Anal. Fertil. Prob.* DSIR Information Series No. 134, Wellington, N.Z., pp. 263–273.

Lee, P.F.W. 1980. Initial observations on the application of the DRIS approach to tea (*Camellia sinensis* L.): Foliar analysis results in South Africa. *Crop Prod.* 9:207–209.

Leech, R.H., and Y.T. Kim. 1981. Foliar analysis and DRIS as a guide to fertilizer amendments in poplar plantations. *For. Chron.* 17–21.

Letzsch, W.S., and M.E. Sumner. 1984. Effect of population size and yield level in selection of diagnosis and recommendation integrated system (DRIS) norms. *Commun. Soil Sci. Plant Anal.* 15:997–1006.

Liebhardt, W.C. 1981. The basic cation saturation ratio concept and lime and potassium recommendations on Delaware's coastal plain soils. *Soil Sci. Soc. Am. J.* 45:544–549.

Lintner, J. 1967. *Fertilizer investigations on Potatoes 1958–1965*. Potash S. Afr., Johannesburg, S. Afr.

McLean, E.O. 1977. Fertilizer and lime recommendations based on soil tests: good, but could they be better? *Commun. Soil Sci. Plant Anal.* 8:441–464.

McLean, E.O., R.C. Hartwig, D.J. Eckert, and G.B. Triplett. 1983. Basic cation saturation ratios as a basis for fertilizing and liming agronomic crops II. Field studies. *Agron. J.* 75:635–576.

Meldal-Johnsen, A., and M.E. Sumner. 1980. Foliar diagnostic norms for potatoes. *J. Plant Nutrit.* 2:569–576.

Melsted, S.W., H.L. Motto, and T.R. Peck. 1969. Critical plant nutrient composition values useful in interpreting plant analysis data. *Agron. J.* 61:17–20.

Meyer, J.H. 1975. Advances in the interpretation of foliar analysis of sugarcane in the South African Sugar Industry. *Proc. S. Afr. Sugar Tech. Assoc.* 49:1–9.

Meyer, J.H. 1981. An evaluation of DRIS based on leaf analysis for sugarcane in South Africa. *Proc. S. Afr. Sugar Tech. Assoc.* 55:1–8.

Munson, R.D., and W.L. Nelson. 1973. Principles and practices in plant analysis. In: L.M. Walsh, and J.D. Beaton (eds.) *Soil Testing and Plant Analysis*. Soil Science Society of America, Madison, WI, pp. 223–248.

Rominger, R.S., D. Smith, and L.A. Peterson. 1975. Changes in elemental concentrations in alfalfa herbage at two soil fertility levels with advance in maturity. *Commun. Soil Sci. Plant Anal.* 6:163–180.

Steel, R.G.D., and J.H. Torrie. 1980. *Principles and Procedures of Statistics: A Biometrical Approach*. McGraw-Hill Book Co., New York.

Sumner, M.E. 1977a. Use of the DRIS system in foliar diagnosis of crops at high yield levels. *Commun. Soil Sci. Plant Anal.* 8:251–268.

Sumner, M.E. 1977b. Preliminary NPK foliar diagnostic norms for wheat. *Commun. Soil Sci. Plant Anal.* 8:149–167.

Sumner, M.E. 1977c. Application of Beaufils' diagnostic indices to maize data published in the literature irrespective of age and conditions. *Plant Soil* 46:359–369.

Sumner, M.E. 1977d. Effect of corn leaf sampled on N, P, K, Ca, and Mg content and calculated DRIS indices. *Commun. Soil Sci. Plant Anal.* 8:269–280.

Sumner, M.E. 1977e. Preliminary N, P, and K foliar diagnostic norms for soybeans. *Agron. J.* 69:226–230.

Sumner, M.E. 1978. Interpretation of nutrient ratios in plant tissue. *Commun. Soil Sci. Plant Anal.* 9:335–345.

Sumner, M.E. 1979. Interpretation of foliar analysis for diagnostic purposes. *Agron. J.* 71:343–348.

Sumner, M.E. 1981. Diagnosing the sulfur requirements of corn and wheat using foliar analysis. *Soil Sci. Soc. Am. J.* 45:87–90.

Sumner, M.E. 1982. *The Diagnosis and Recommendation Integrated System (DRIS).* Soil/Plant Analysts Workshop, Council on Soil Testing and Plant Analysis, Anaheim, CA.

Sumner, M.E. 1983. The Diagnosis and Recommendation Integrated System (DRIS) as applied to sugarcane. *Inter-American Sugarcane Seminar Soil Fertil. and Mgmt.* Miami 1:7–19.

Sumner, M.E. 1985. The Diagnosis and Recommendation Integrated System (DRIS) as a guide to orchard fertilization. Food & Fertil. Technology Center, Ext. Bull. 231, FFTC/ASPAC, Taipei, Taiwan, R.O.C.

Sumner, M.E., and P.M.W. Farina. 1986. Phosphorus interactions with other nutrients and lime in field cropping systems. In: *Advances in Soil Science,* Vol. V, B.A. Stewart (ed.), Springer-Verlag, New York, pp. 201–236.

Terman, G.L. 1974. Yield nutrient concentration relationships in maize. *Proc. 7th Int. Coll. Plant Anal. Fertil. Prob.* 2:447–458.

Tserling, V.V. 1974. Principal results of investigations on nutritive diagnosis of major crops in different natural zones of the U.S.S.R. In: J. Wehrmann (ed.) *Proc. 7th Int. Coll. Plant Anal. Fert. Prob.* German Soc. of Plant Nutrition, Hanover, GDR, pp. 459–472.

Ulrich, A., and F.J. Hills. 1967. Principles and practices of plant analysis. In: M. Stelly (ed.) *Soil Testing and Plant Analysis.* Part II. *Plant Analysis.* Soil Science Society of America, Spec. Pub. 2. Madison, WI, pp. 11–24.

Walworth, J.L., and M.E. Sumner. 1986. Foliar diagnosis—A review. In: *Advances in Plant Nutrition* Vol. III. B.P. Tinker (ed.) Elsevier, New York (In press).

Walworth, J.L., W.S. Letzsch, and M.E. Sumner. 1986a. Use of boundary lines in establishing diagnostic norms. *Soil Sci. Soc. Am. J.* 50:123–128.

Walworth, J.L., M.E. Sumner, R.A. Isaac, and C.O. Plank. 1986b. Preliminary DRIS norms for alfalfa. *Agron. J.* 78:1046–1052.

# Soil-Related Nutritional Problem Areas for Grazing Animals*

Joe Kubota, Ross M. Welch, and Darrell Van Campen

I. Introduction ........................................................ 189
II. General Considerations of Trace Element Problems in Plants and Animals ........................................................ 190
A. Group I Elements ........................................................ 191
B. Group II Elements ........................................................ 193
C. Group III Elements ........................................................ 193
D. Group IV Elements ........................................................ 194
III. Geographical Distribution of Nutritional Problem Areas for Grazing Animals ........................................................ 194
A. Cobalt ........................................................ 196
B. Molybdenum ........................................................ 198
C. Copper ........................................................ 201
D. Selenium ........................................................ 203
IV. General Discussion ........................................................ 205
V. Future Outlook ........................................................ 211
References ........................................................ 212

## I. Introduction

Research to correct trace element deficiencies and toxicities in plants has resulted in increased production of many food and forage plants in the United States and in other parts of the world. The increases in forage crop production have contributed to overall increases in animal production. However, increased animal production has not always paralleled increased production of forage crops. Such disparities arise when meeting the nu-

*Contribution from the US Plant, Soil and Nutrition Laboratory, Tower Road, Ithaca, New York 14853 and the Department of Agronomy, Cornell University, Ithaca, New York 14853.

tritional requirements and mineral tolerances of the plant fails to satisfy the nutritional needs of the animals that are eating the plants. In Australia, use of trace elements has transformed land destined as "trace element deserts" into productive grazing land (Anderson and Underwood, 1959). However, some of these improved pastures were less productive than others when productivity was assessed in terms of animal production. Recognizing the trace element requirements of grazing animals is essential when forage plants are grown primarily for animal feed.

In the United States, there are naturally occurring areas where grazing animals have enough feed to meet energy and protein needs but fail to gain weight and grow normally because of a nutritional deficiency or toxicity of a trace element. The geographical distribution of such nutritional problem areas and the soil and plant factors associated with them have been investigated in an effort to raise overall agricultural production through improved crop and animal management practices. The present discussion focuses principally on trace elements, because deficiencies and toxicities of trace elements cause nutritional problems in grazing animals far out of proportion to those commonly found in soils and plants. Information based on experimental animals and people is given where applicable.

## II. General Considerations of Trace Element Problems in Plants and Animals

A list of trace element deficiencies and toxicities in plants and animals is presented in Table 1. Fewer trace elements are considered here than were considered by Nielsen (1984) and Welch (1985) in their comprehensive reviews. Problems of deficiencies and toxicities in plants are based largely on observations of field-grown plants. Those problems in plants and animals observed mainly under experimental conditions (expt.) are indicated. Among the trace elements considered (Table 1), deficiencies exceed toxicities among both plants and animals. Differences in trace element requirements between plants and animals reflect, in part, the current state of knowledge and interest in the two fields of study. They may also reflect differences in how essentiality of a trace element is defined and applied to plants and to animals (Welch, 1985). All species of plants and animals are not equally sensitive to any given trace element, so the deficiencies or toxicities outlined in Table 1 probably reflect such a bias. For example, vegetables and fruit trees are most often associated with deficiencies of trace elements (Berger, 1962). The essentiality to animals of some of the more recently recognized trace elements often is based on observed effects on laboratory animals raised under controlled experimental conditions (Nielsen, 1984).

**Table 1.** Nutritional problems of deficiencies and toxicities in plants and animals associated with trace elements[a]

| | Element | Plant[b] | Animal[b] |
|---|---|---|---|
| I. | Required by most plants and animals | | |
| | B | D/T | D/T |
| | Mo | D | T (molybdenosis) |
| | Cu | D/T | D (Mo induced) |
| | Zn | D/T | D |
| | Fe | D/T | D |
| | Mn | D/T | D |
| II. | Required by animals and some species of plants | | |
| | Co | D | D |
| | Ni | D/T | D (expt) |
| | Si | D | D (expt) |
| III. | Required by animals but have some beneficial effects on plants | | |
| | Se | (expt) | D (Se-responsive diseases) |
| | | | T (selenosis) |
| | I | (expt) | D (goiter) |
| | Cr | (expt) | D |
| IV. | Possibly required by animals but not by plants | | |
| | Sn | (—) | D (expt) |
| | Cd | (—) | D/T (expt) |
| | Pb | (—) | D/T (expt) |
| | | | T (environmental pollutant) |
| | F | (—) | T |

[a] Adapted from Nielsen (1984) and Welch (1985).
[b] D, deficiency; T, toxicity; (expt), experimental; (—), not established or known.

## A. Group I Elements

In Table 1, the six trace elements in group I are universally recognized to be essential to both plants and animals. Of these, B and Mo can be toxic to both plants and animals. In Russia, high concentrations of B in drinking water and in feed plants (Underwood, 1971) have been associated with a nutritional problem in sheep. High B concentrations (200 to 400 μg/g) have been found in forage and native range plants and in waters of some thermal springs (0.1 to 3.0 μg $ml^{-1}$) in Nevada (Kubota, unpublished data, 1983). Plants having high concentrations of B grew in areas draining the high-B thermal spring waters. High B concentrations also were found in plants from soils that had accumulated B salts over time. Both the water and the plants are important dietary sources of B for livestock that graze areas that receive drainage from B-rich water from thermal springs. Water and lush vegetation are available to livestock in an otherwise dryland area. However, specific nutritional problems associated with dietary B intake by cattle in Nevada have not been demonstrated.

In the case of Mo, animal problems reflect dietary imbalance between Mo and Cu (Mo-induced Cu deficiency) in feed plants. Cattle and sheep afflicted with molybdenosis show symptoms of Cu deficiency and respond to Cu supplements. Forage concentrations of Mo in the range of 10 to 20 $\mu g\ g^{-1}$ or more are associated with Mo-induced Cu deficiency in cattle. Animal problems associated with excess plant Mo concentrations are more prevalent than those arising from low plant Cu concentrations. Uncomplicated deficiencies of Cu, independent of Mo, are rarely observed under field conditions. Legumes accumulate Mo and are useful indicators of the Mo supplying power of soils; however, they also enhance the movement of Mo from soil to plants to animals.

Zinc deficiency is more likely to be a nutritional problem than is Zn excess in either plants or animals. Zinc toxicity has been observed only under special conditions; i.e., it has been observed in plants grown in an organic soil area, underlain by Zn sulfide deposits, that was drained and then cropped (Staker, 1943). Zinc deficiency in plants has been observed under a wide range of conditions and in a number of species. Zinc deficiency is common in certain crops grown on alkaline soils. Citrus trees are often fertilized with Zn. Deficiencies are frequently observed when fields have been graded for surface irrigation. Where topsoil has been removed during the leveling process, crops such as corn (*Zea mays* L.) and soybeans (*Glycine max* L. Merr.) may become Zn deficient and be very stunted. Zinc deficiency has been observed in poultry and swine fed diets based on soy protein. Deficiencies in grazing animals are less common; however this may be, in part, because there is no reliable method for detecting marginal Zn deficiencies. In one series of feeding trials in the Western United States, range cattle responded to supplemental Zn with increased weight gains (Mayland *et al.*, 1980).

Iron deficiency in plants usually occurs on alkaline or very sandy soils, although some plants develop Fe deficiency on many different soils. Iron deficiency in plants is rarely, if ever, the result of low total Fe in the soil but is the result of inadequate available iron levels. In animals, Fe deficiency is most prevalent in baby pigs raised on concrete floors and in certain human population groups. It has rarely been reported in grazing animals. Iron toxicity is rarely a problem in either plants or animals. In humans, it is usually associated with metabolic disorders rather than with excessive intakes.

Manganese is required for both plants and animals. Manganese deficiency has been observed in small grains grown on alkaline soils. In acid soils, Mn is much more soluble and there are a number of reports of Mn toxicity of plants under acid soil conditions. In animals, Mn deficiency has been reported most often in poultry and results in bone abnormalities, particularly in the long bones of the legs. Few cases of Mn deficiency have been reported in pigs, cattle, and sheep; however, there have been cases of reproductive problems and bone abnormalities that were attributed to Mn deficiency. Manganese toxicity is not generally considered a problem in animals or in people.

### B. Group II Elements

Cobalt deficiency has been a more widespread problem in ruminant animals than in plants. Among plants, legumes require Co for N fixation. Discoveries in Australia and New Zealand that ruminant animals respond to Co supplementation have led to correction of Co deficiencies in the United States. Prior to its association with Co deficiency, these animal disorders were known in New Hampshire as "Chocorua's curse" and "Albany ail," in Massachusetts as "neck's ail," and in Michigan as "Grand Traverse disease." The other two trace elements considered with Co (group II) are Ni and Si. The essentiality of Ni for animals was demonstrated by Nielsen (1974), and its essentiality for legumes was subsequently demonstrated with the soybean plant (Eskew *et al.*, 1983). Such grasses as rice (*Oryza sativa* L.) and sugar cane (*Saccharum officinarum* L.) have responded to supplemental Si and the essentiality of Si for animals was first demonstrated in chicks by Carlisle (1974). Deficiencies of Ni or Si, in field-grown crops or in grazing animals, have never been reported.

### C. Group III Elements

Selenium, I, and Cr have a greater impact on animals than on plants. Selenium probably affects a larger number of animal species than any other trace element considered in this report; i.e., it affects birds and poultry as well as ruminant and nonruminant animals, including humans (Thacker, 1961; Welch and House, 1984). Both a deficiency of Se (Se-responsive diseases) and toxicity (selenosis) are important nutritional problems among a wide range of animals. The Se-deficient areas are extensive compared with the localized occurrence of Se toxic areas. Over the range from deficiencies to toxicities, plant Se concentrations vary from less than 0.01 $\mu g\ g^{-1}$ to several thousand micrograms of Se per gram. In animals, Se deficiency problems are encountered when plant concentration is in the range of 0.05 $\mu g\ g^{-1}$ or less and, toxicities generally occur at concentrations of 4 to 5 $\mu g\ g^{-1}$ or more.

An association of I deficiency with goiter is well established, but a geographical pattern of goiter incidence with soils is poor. An association of the disease incidence with proximity to sea water (i.e., the further from sea water, the greater the incidence) seems evident in humans but this observation may be confounded by effects of diet as well (Kubota and Allaway, 1972).

Chromium is essential for optimum glucose utilization by animals and man, but the requirement of Cr to be present in a specific form (glucose tolerance factor) to be biologically effective (Mertz, 1969) currently limits the value of Cr concentrations determined in plants.

### D. Group IV Elements

Tin, Pb, and Cd do not have well-established biochemical roles in plants but ultratrace concentrations are considered to be beneficial (experimental) to animals (Nielsen, 1984). Cadmium and Pb in particular have been studied extensively because of their toxic effects on humans. Toxicities of Cd and Pb in livestock and people have been associated primarily with industrial pollution. Under natural conditions neither soils, plants, nor animals, individually or collectively, appear to serve as absolute barriers to the transfer of Cd in the food chain of man.

## III. Geographical Distribution of Nutritional Problem Areas for Grazing Animals

The nature of the nutritional problem areas for grazing animals in the United States is illustrated from the viewpoint of four trace elements: Co, Mo, Cu, and Se. Copper is included because of its interaction with Mo. Specific problem areas are tied by trace element concentrations in plants to observed nutritional deficiencies or toxicities in animals. Information about soils and soil maps is used to depict the geographical distribution of the problem areas after soil characteristics that are common to the problem areas at the various locations are identified.

A summary of soil and plant factors associated with Co deficiencies, Mo toxicities, and Se deficiencies and toxicities in the animal is presented in Table 2. Plant concentrations are used to assess the trace element supplying power of the soils, which is influenced by soil parent materials (soil capacity factor) and soil weathering (intensity factor).

An influence of the geological origin of soil parent material is usually evident in areas where problems of deficiencies or toxicities in animals are acute. Information about the rock sources of trace elements is available from geochemical reports (Rankama and Sahama, 1950; Goldschmidt, 1954). Because plants in many areas grow on soils formed in unconsolidated surficial deposits, the effects of the possible redistribution of the trace elements through mixing and resorting processes associated with transport and deposition of the surficial materials are considered. The use of taxonomic units of soils (Soil Conservation Service, 1975) in the selection and location of sampling sites provides a basis to apply soil classificational units and soil maps of suitable scales to depict the distribution of the nutritional problem areas.

Effects of plant species (i.e., genetic variability between species) on the trace element concentrations in plant organs are important consid-

**Table 2.** Role of soils, plants, and soil parent materials associated with nutritional problem areas in the United States

| Trace element | Problem | Significant plant concentration levels ($\mu g\ g^{-1}$) | Role | | |
|---|---|---|---|---|---|
| | | | Plants | Soil | Soil parent material |
| Cobalt | Deficiency | $\leq$ 0.04 to 0.07 | Grasses: poor absorbers of Co<br>Black Gum: accumulates Co | Acid, wet sands, and loamy sand (Spodosol) | Coastal Plain deposits<br>Glacial drift, local rock sources |
| Molybdenum | Toxicity | 10 to 20 or more | Legumes: accumulate Mo | Neutral to alkaline, wet soils<br>Peats and muck | Alluvium of granitic rock and shale on alluvial fans and floodplains of small streams |
| Selenium | Deficiency | $\leq$ 0.05 | Effects of species not pronounced at deficiency levels | Acid | Parent rock generally older than Cretaceous |
| | Toxicity | 4 to 5 or more | Pronounced effect of species;<br>Se accumulator plants | Alkaline, calcareous | Seleniferous, sedimentary rocks, mostly shale |

erations, so steps are taken during the course of an investigation to collect all plant samples by species from nearly all of the sampling sites. The general availability of legume–grass mixtures in hayfields and pastures provides almost unlimited sampling sites for a wide range of soils. Following preliminary assessment of plant concentrations, the data for legumes may be treated separately from those of the grasses (Co, Cu, and Mo) or they may be combined as in the case of Se. Effects of plant species are almost negligible at Co and Se concentrations at or below deficiency levels for animals, but species differences are often most evident and pronounced at concentrations toxic to the animals.

### A. Cobalt

Cobalt deficiency has been a nutritional problem in animals in parts of the Northeastern United States since early settlers cleared forested land to grow agricultural crops. In the Eastern United States, the low-Co areas are associated with acid, sandy Spodosols (Figure 1). In northern New Hampshire and southern Maine, the low-Co area largely coincides with the areas mantled by glacial drift derived from granite of the White Mountains (Figure 2). The sandy soils have less Co than do similar sandy soils outside of the area, which are soils formed in glacial drift from mixed rock sources (Kubota, 1964). On the lower Atlantic Coastal Plain, the low-Co areas are coextensive with the distribution of Humaquods of sand and loamy sand textures. The Humaquods have the least Co of any soils in the United States (Kubota, 1972). The low Co content of the C horizons of these soils suggested that the coastal plain deposit in which Humaquods formed was a poor soil parent material, containing inadequate amounts of available Co. The distributional pattern of Co (total and extractable) with soil depth indicates that an appreciable amount of Co has been leached below the rooting depth of most plants. The leaching of Co appears to have occurred as organic complexes of Co, forms that should also be available to plants. The low Co concentration in plants grown on Humaquods probably is accentuated by an organic pan that restricts the rooting depth of most plants.

In the Southeastern United States, an association of Humaquods with Co deficiency in animals was also evident when the Co concentration in leaves of black gum (*Nyssa sylvatica biflora* (Walt.) Sarge.), a Co accumulator plant, was used to assess the Co-supplying power of the soils. Cobalt concentrations in samples of black gum leaves collected from North Carolina to Texas ranged from 1 to nearly 1000 $\mu g\ g^{-1}$ (Kubota and Lazar, 1958; Kubota *et al.*, 1960). Within this broad concentration range, the samples from the sandy Humaquods of North Carolina and South Carolina, areas where Co deficiency has been critical in cattle, had the least amount of Co (1 to 5 $\mu g\ g^{-1}$). By comparison, samples of grasses (*Andropogon virginicus* L.) collected from North Carolina to Texas from a number of

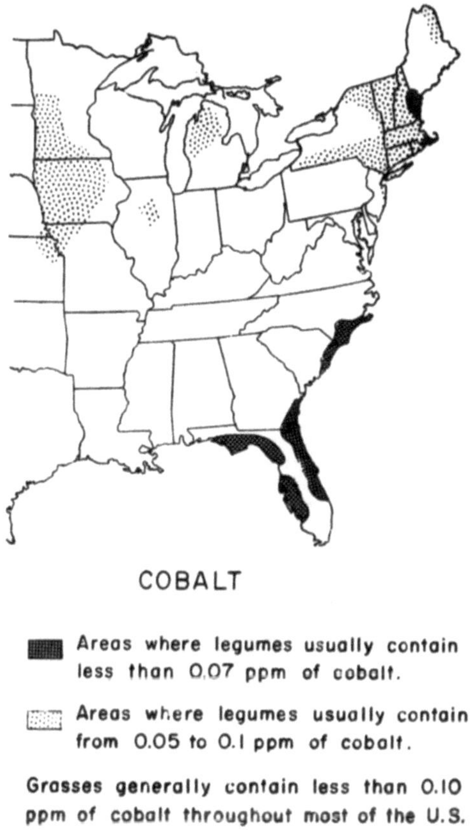


**Figure 1.** Geographical distribution of Co-deficient areas in the eastern United States (ppm = 1 $\mu g/g^{-1}$). From Kubota and Allaway, 1972, by permission Soil Science Society of America.

the same sites had from 0.05 to 0.63 μg of Co per gram. The black gum plant, although useful as an indicator plant of soil Co, does not seem to be a plant source of Co for grazing animals.

An association of Co deficiency in animals with soil parent material was also evident in New Zealand. On these islands, Co deficiency is coextensive with areas of soils formed in deposits of Taupo ash and Karahoa pumice (Andrews, 1956). Volcanic ash and pumice also mantle broad areas in Oregon where grazing is important, but the soils formed in the ash deposits here support forage plants with adequate Co levels for cattle. The ash deposits in New Zealand are mostly andesitic, whereas those in Oregon are basaltic.

In the New England states, grasses are poor plant sources of Co for cattle (Figure 3). At mean Co concentration levels of 0.04 $\mu g\ g^{-1}$, differences between species of grasses are not evident. Because grasses tend to have uniformly low amounts of Co, a dominance of grasses in pastures and hayfields in the New England states could mask any association between soils and the incidence of Co deficiency in livestock. More likely,

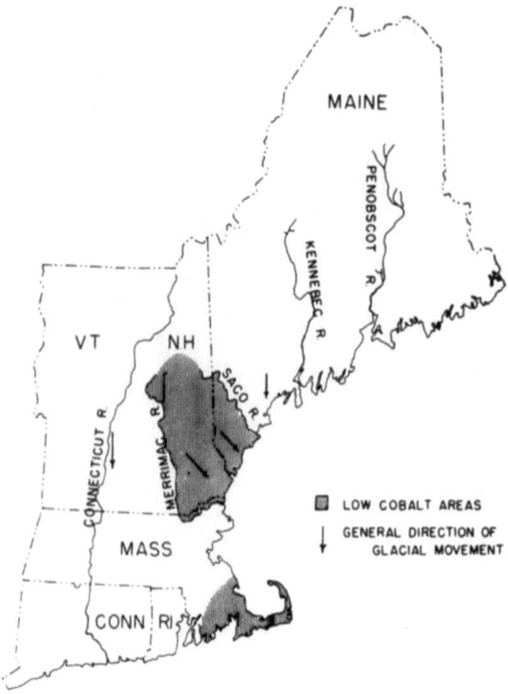


**Figure 2.** Distribution of the low-Co areas in New England. From Kubota, 1964, by permission Soil Science Society of America.

the dominance of grass forages would tend to exaggerate the geographical extent of the low-Co areas.

### B. Molybdenum

Cattle were observed to weaken rapidly when they grazed in parts of the eastern footslopes of the Sierra Nevadas in Nevada. The poor condition of the cattle was attributed to "alkali disease" and the forage was cut and fed as hay. Subsequent investigations revealed the cause to be molybdenosis, resulting from an excess of Mo in forage plants (Kubota *et al.*, 1961). The problem areas were found to be confined principally to wet midfan and toe positions of some granitic alluvial fans. When Mo concentrations were compared in similar plants from similar wet soils on five parallel alluvial fans, the Mo concentrations were more uniform in plant samples from wet soils of a given alluvial fan than from similar species of plants from similar soils on different alluvial fans. Within one alluvial fan, the soil Mo tended to increase downslope with changes in soils from the top of the fan (gravely, loamy soil), to midfan (sandy loam), to the toe of the fan (loam). These changes in texture of the soils downslope paralleled increases in the soils' organic matter content (Kubota, 1977).

The variations in plant Mo content between alluvial fans were also ob-

**Figure 3.** Frequency distribution of Co concentration in grasses from widely differing soils of New England (ppm = 1 $\mu g/g^{-1}$). From Kubota, 1964, by permission Soil Science Society of America.

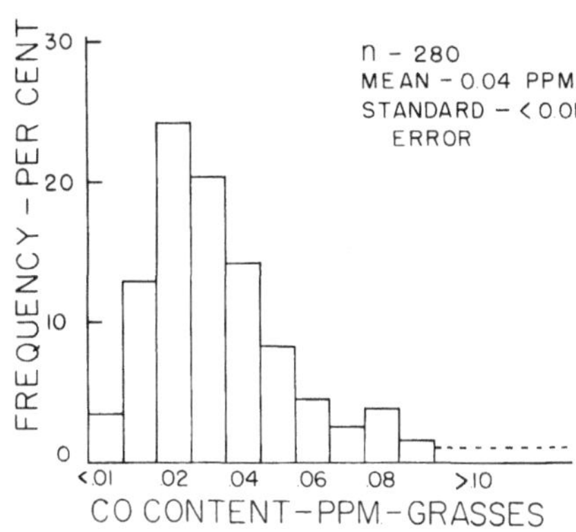


served on floodplains of parallel, tributary streams of the Powder River in Baker Valley, Oregon (Kubota *et al.*, 1967b). Here, the wet soils of two stream floodplains had appreciably more Mo than did the similar wet soils of the other stream floodplains (Figure 4). The differences in soil Mo concentrations were reflected in plant Mo concentrations (Figure 5). The soil–plant relationship noted indicated that the mineral carriers of Mo were the same in the soils of Baker Valley.

Unlike the tributary stream system, the Mo concentration was uniformly low in soils and plants from the Powder River floodplain and was comparable to Mo concentrations in plant samples from the floodplains of the Columbia River and the Williamette River in Oregon. Alluvial deposits of large rivers are derived from many stream sources. Consequently, excessive amounts of Mo carried by any one tributary stream tends to be diluted with deposits of other streams. Thus, molybdenosis is not a problem of floodplains of large rivers, even though parts of the floodplains may be wet. Similar variability in soil and plant Mo concentrations associated with small stream systems in the intermountain valleys in Nevada and Oregon was observed in other western states (Kubota, 1975).

Granite is the rock source of soil parent material that is associated with most of the Mo-toxic areas in Nevada, Oregon, Washington, Idaho, and California. Shale is also an important rock source of soil Mo. Soils formed in or influenced by shale of the Lower Lias formation in England have been associated with the teart areas of high Mo, where Mo toxicity in farm animals was first diagnosed (Ferguson *et al.*, 1943). Toxic levels of Mo have been observed in plants grown on shale-derived soils in Colorado, but the Mo concentrations of plants grown on these soils have been relatively low when compared with Mo concentrations of 100 to 400 $\mu g\ g^{-1}$ in plants grown on soils formed in granitic alluvium. The Mo in soils derived from shale does not seem to be as readily available to plants as is the Mo in soils derived from granitic rock sources (Kubota, 1977). One of the highest soil Mo concentrations (31.5 $\mu g\ g^{-1}$) was found in a soil of

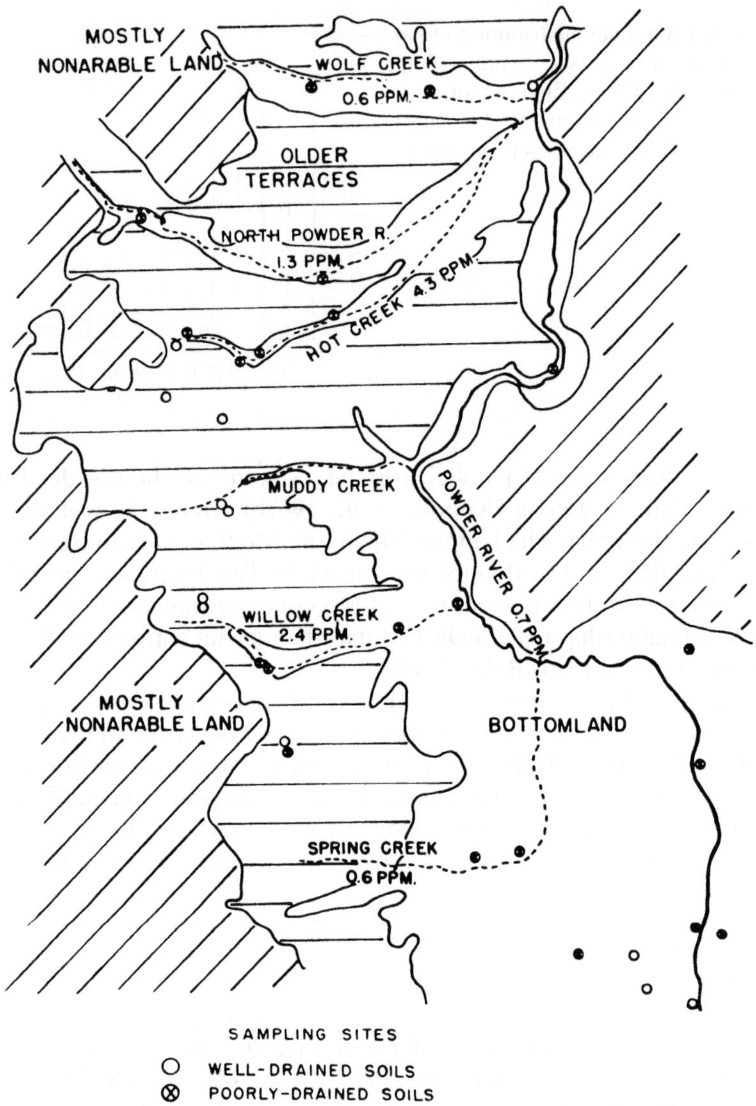


**Figure 4.** Molydenum content of soils in Baker Valley, Oregon, in relation to tributary stream sources (ppm = 1 μg/ g$^{-1}$). From Kubota *et al.*, 1967b, by permission Soil Science Society of America.

Virginia that formed in weathered products of interbedded shale and limestone (Robinson and Edgington, 1954). This well-drained, acid soil has a large amount of Mo but has not been implicated with molybdenosis in cattle. The median soil Mo concentration in a wide range of United States soils is slightly more than 1 μg g$^{-1}$ (Kubota, 1977).

The availability of soil Mo to plants is high in most of the neutral to

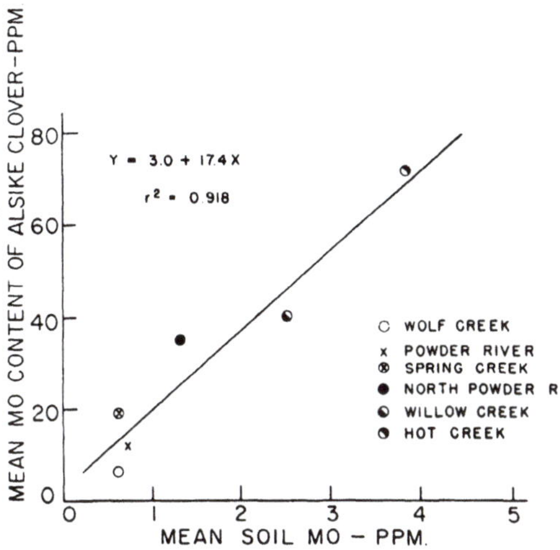


**Figure 5.** Molybdenum concentration in legumes in relation to soil molybdenum (total) in A horizons of wet soils, Baker Valley, Oregon (ppm = 1 μg/ g$^{-1}$). From Kubota *et al.*, 1967b, by permission Soil Science Society of America

alkaline soils in the western states. Liming an acid, Mo-rich (31.5 μg g$^{-1}$) soil of shale rock origin increased plant concentrations, but the effects of liming were small (Robinson *et al.*, 1951) when compared to effects of soil wetness on increased plant concentrations (Kubota *et al.*, 1963).

Effects of rock sources of soil parent materials combined with soil reaction contribute to a pattern of gradually decreasing Mo concentration in plants from the west to the east United States (Figure 6). The map is based on Mo concentration in legumes sampled across the United States from a wide range of soils that differ in drainage, soil pH, and soil parent materials. Generalized areas where molybdenosis occurs are shown against a background of plant Mo concentrations of nonproblem areas. The problem areas in the western states generally are not extensive, because they occur only on wet parts of some alluvial fans and floodplains of small streams that project out onto the floor of the intermountain valleys and basins.

Molybdenosis also is a problem in two broad areas of peat and muck soils in the United States. The Delta area in California, at the confluence of the San Joaquin and Sacramento Rivers, is a broad organic soil area where molybdenosis is a problem. This and the Florida Everglades, represent fairly extensive areas where wet organic soils have been drained and then cropped. In Florida, marl that underlies peats and mucks is a factor contributing to the growth of high-Mo forage plants.

## C. Copper

Because excess Mo concentrations in forage plants induce Cu deficiencies in grazing animals and the afflicted animals respond to Cu supplements, Cu concentrations were determined routinely in all plant samples collected

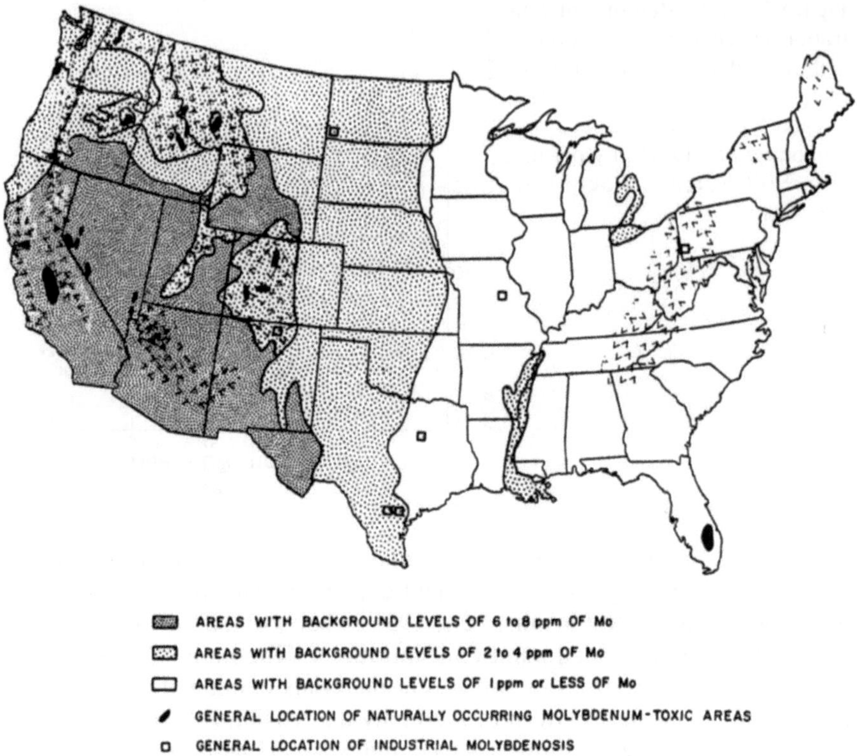


**Figure 6.** Generalized regional pattern of molybdenum concentration in legumes of the United States (ppm = 1 $\mu g/g^{-1}$). From Kubota, 1977, by courtesy Marcel Dekker, Inc.

to study Mo. Additionally, Cu concentrations were determined in plant samples collected for the study of other trace elements (i.e., Co, Se, Cr, and Cd) (Kubota and Lazar, 1960; Kubota *et al.*, 1960, 1967a; Kubota, 1968). The Cu values obtained thus provide representative data for forage plants from most agricultural regions of the United States. In all cases, the Cu values included in the summary were obtained after separation of Cu by dithizone extraction of plant material digestates.

In general, the range of Cu concentrations in forage plants is narrow and there was no pronounced effect of soils on Cu concentrations in either legumes or grasses (Kubota, 1983a). The distribution pattern of Cu in legumes from different regions of the United States is presented in Figure 7. The median and range of concentrations were nearly the same in legumes from the broad region of acid soils in the East and in legumes from dominantly alkaline and calcareous soils in the West. Copper concentrations were low in legumes from two dissimilar soils areas (35 percent or more of the legumes had 6 $\mu g$ g-1 or less of Cu): (a) the low Co areas in the eastern

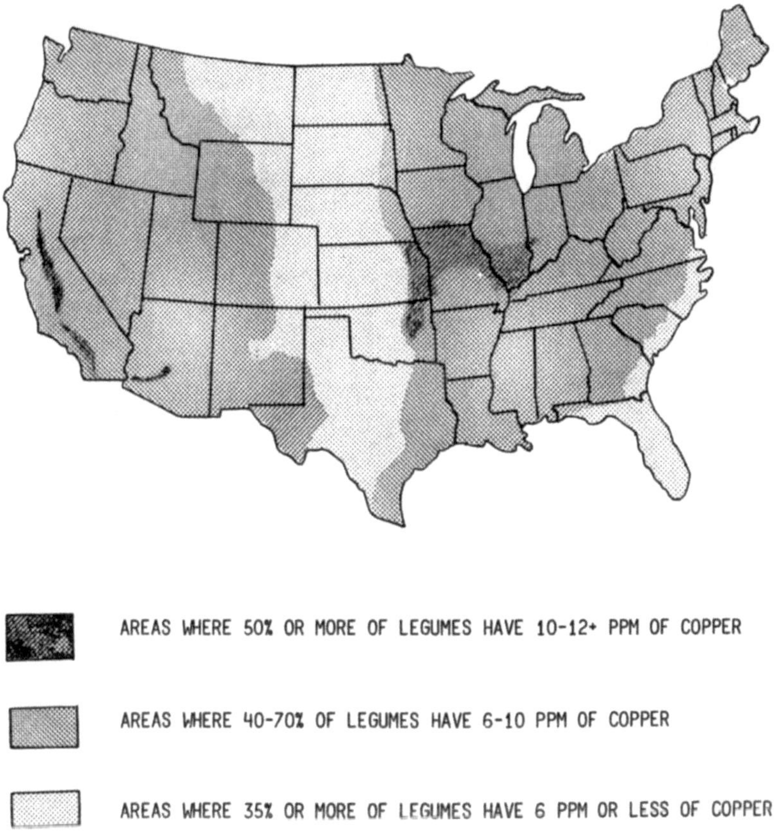


**Figure 7.** Generalized distribution of copper concentration in legumes of the United States (ppm = 1 μg/ $g^{-1}$). From Kubota, 1983a, by permission Amer. Society of Agronomy.

United States, and (b) the wheat belt in the central United States. Use of Mo/Cu ratio has been proposed to identify potential Mo-toxic areas for grazing animals in the United States. The generally uniform level of Cu in plants limits the usefulness of such a ratio in the understanding of the causal factors that contribute to molybdenosis in animals. Incidences of molybdenosis appear to be more strongly associated with marked increases in plant Mo and less so with changes in Cu levels. Soil wetness had little or no effect on plant concentrations of Cu (Kubota *et al.*, 1963).

### D. Selenium

The broad regional pattern of plant Se concentrations in the United States is presented in Figure 8. The central United States extending from Canada to the Gulf is in the adequate range for Se. This is also the region where

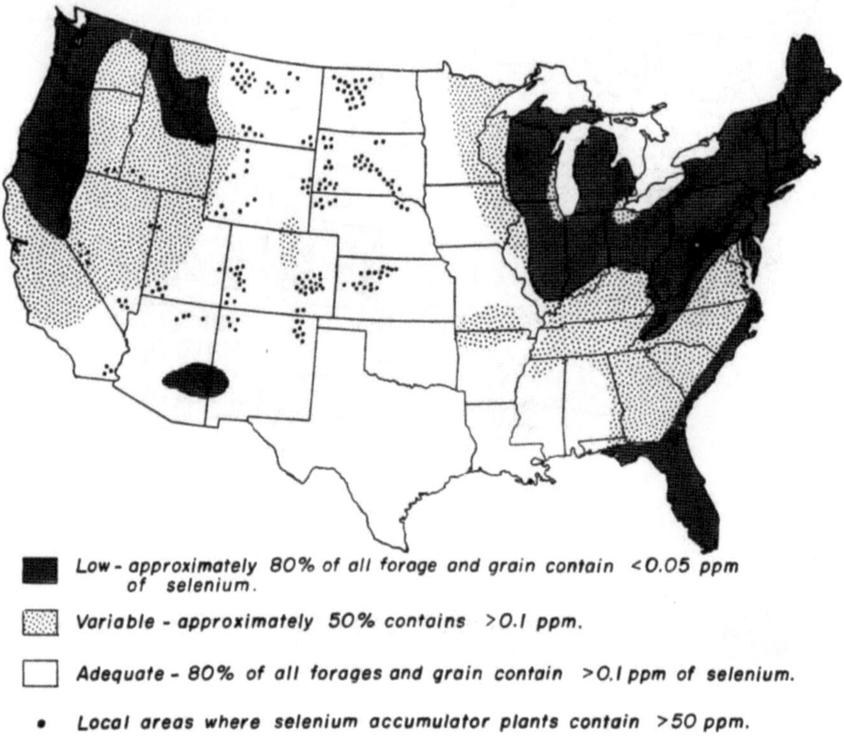


**Figure 8.** Geographical distribution of low-, variable-, and adequate-Se areas in the United States (ppm = 1 μg/g). From Kubota and Allaway, 1972, by permission Soil Science Society of America.

Se toxicity is a nutritional problem locally where Se accumulator plants with toxic levels of Se grow on seleniferous soils. The geographic pattern that emerges largely reflects the influence of Cretaceous and Tertiary sedimentary rocks that locally are rich rock sources of Se. The broad geographic extent of the Se adequate area strongly reflects the redistribution of the Se on the landscape from its rock sources to Se in surficial deposits, glacial drift, loess, and alluvium. In the northern parts of this area the soils are calcareous, but they are acid in northern Missouri and southwards to Mississippi. The distribution pattern of plant Se reflects the dominant role that the Missouri River and the Mississippi River drainage systems have played in redistribution of Se in surfical deposits in the central United States. Toxic concentrations of Se have not been observed in plants grown on acid soils. The Se-toxic areas in the north-central and Rocky Mountain states have calcareous soils that have formed in weathered seleniferous rocks or in materials derived from them. The selenate forms of Se present in such soils are largely available to plants. Shales of the Niobara and Pierre formations of Cretaceous age are the best known of the seleniferous rocks. The distribution of the Se-toxic areas, however, is spotty and does not uniformly cover the landscape where Cretaceous and Tertiary shale occur, because the rocks are not uniformly seleniferous throughout their geographical extent (Lakin, 1961a). For example, the median Se concentration of 474 samples of rocks collected from 69 strata of the Niobara and the Pierre formations was 3.0 $\mu g\ g^{-1}$ (see summary by Kubota, 1972). The Se concentrations also varied widely from platelet to platelet within the layered rocks. The presence of selenate forms of Se together with the growing of Se accumulator plants, such as *Astragalus bisulcatus* Hook. and *Stanleya bipinnata* (Pursh.) Britt. (Beeson, 1961), magnifies the effects of the seleniferous soils. Most cases of acute Se poisoning in animals have been associated with livestock eating Se-accumlator plants.

Low-Se areas occupy parts of the Pacific Northwest, the Atlantic Coastal Plain, and much of the northern part of the United States from the Lake States to New England. A feature common to the areas of low Se is the growing of forage plants on acid soils. In the Pacific Northwest, the acid soils are formed in or influenced by volcanic ash deposits, but elsewhere the acid soils have formed in unconsolidated surficial deposits that lack well-recognized rock sources of Se (Muth and Allaway, 1963; Kubota *et al.*, 1967a). The plants in these areas have small amounts of Se and the incidence of Se-responsive diseases is high (Muth and Allaway, 1963).

The variable-Se areas are transitional zones where forage plants do not consistently have either low or adequate amounts of Se. In parts of Oregon, Nevada, and states contiguous to them, plants with low Se that grow in the variable-Se area appear to be from soils formed in or influenced by volcanic ash. In many places these soils are calcareous and alkaline, but the plants from these soils tend to contain low Se levels for animals. The variable-Se area in the eastern United States includes soils formed in weathered products of the underlying sedimentary rocks, mostly limestone. In the western parts of Illinois, Kentucky, and Tennessee, loess cap that mantles soils in parts of these states may contribute to the variability of Se concentrations in plants. The variable-Se area in parts of Minnesota and Wisconsin is a soil transition zone where a calcareous subsoil horizon is present in some soils and is absent from others. The uptake by plants of subsoil Se (Olson *et al.*, 1942) may be a soil–plant factor that contributes to variability of plant Se concentrations in this part of the country.

## IV. General Discussion

Several considerations are relevant when studies are proposed to identify and characterize trace element nutritional problem areas for grazing animals. Among these are (1) the geographical scope of the proposed study, (2) the availability of field support to locate sample sites and to collect samples, and (3) the adequacy of laboratory support for trace element analysis. Each contributes to how well the characteristics of the problem areas can be defined so that their distribution pattern on the landscape can be portrayed accurately.

A grid system of sampling has been used widely in geochemical studies to locate geological anamolous areas with mineral resource potential. This system has been applied to investigate the geochemical environment (Shacklette *et al.*, 1971; Howarth and Thornton, 1983) and agricultural problems (Thornton, 1983). Adaptations have been made mainly to obtain background levels of various trace elements in surficial materials of possible importance to animals and man. In this approach, samples are collected in a preset manner from a gridded landscape, usually in a pattern of 1-mi (1.6-km) squares and analyzed by a common multielement analytical procedure. The data are programmed for computer analysis so that they can be reproduced in map form. The overall procedure from sampling to data processing is designed to handle large number of samples and data. Some inherent limitations in the grid system should be recognized when considerations are made regarding use of this system. One consideration is the possibility that some problem areas, such as a Mo-toxic alluvial fan, may be missed in sampling. Another consideration is the possibility that some important differences in trace element concentration may be overlooked because of a lack of sensitivity often inherent in multielement analytical procedures.

A problem-oriented soil–plant approach directed toward solving specific soil-related problems has been developed and used over time at the US Plant, Soil and Nutrition Laboratory, Ithaca, NY, in cooperation with the Soil Conservation Service, USDA. Sample site selection and location are made with the assistance of the field soil scientists of the Soil Conservation Service, and laboratory analysis and interpretations of the data are made in collaboration with chemists, plant physiologists, and animal scientists of the laboratory staff. Concepts of the trace element supplying power of soils and their validity as they apply to soil systems with different rock sources of trace elements and different weathering patterns that affect the release of the soil trace elements are evaluated. In practice, conceptual models of problem soils are developed and used to locate sampling sites of real soils in the field (Kubota, 1972). Nonproblem soils that serve as controls are treated in the same manner. This approach is well adapted to studying the soil-related nutritional problem areas in the United States. For example, there are 613 million hectares (1513 million acres) of privately owned land in the United States, of which 25% is in rangeland and pasture, about 25% is in forest, and about 27% is in crops, including grain crops. Some systematic means to select and locate sampling sites for investigation is desirable and essential.

At each sampling site, the soils are first described and classified according to soil taxonomy (Soil Conservation Service, 1975). The plant samples are then taken by species and a sample is taken of the soil, usually the topsoil, on which the plants have grown. Indicator plants of specific trace elements, if known, are sought and collected where available. A separate sample set is collected for each study and laboratory methods are developed or modified to meet the specific needs of the study. Accepted levels of trace element concentrations in feed and forage plants, associated with specific nutritional deficiencies or toxicities (Underwood, 1971), are used as a basis to summarize the data. When problem soils are identified, soil maps of appropriate scale are used to prepare maps of the problem area.

The soil–plant approach has been used to develop the Co map of the New England states (Kubota, 1964) and the Mo map of the western states (Kubota, 1975) as well as the national maps of Co (Kubota, 1968), Mo (Kubota, 1977), Cu (Kubota, 1983a), and Se (Kubota *et al.*, 1967a). The basic concepts were also used to investigate two other soil-related problems in grazing animals. In one case, some thermal springs (capacity factor) were considered to be likely sources of $F^-$ for $F^-$-toxic plants (Kubota *et al.*, 1982). The role of soils (intensity factor) in this case was minimal because of $F^-$ fixation in calcareous soils (National Academy of Sciences, 1971).

In another study, the Mg concentration in grasses from soils formed in or influenced by dolomite (capacity factor) was compared with Mg in grasses from soils formed in or influenced by calcitic limestone (Kubota *et al.*, 1980). The sampling focused on these two kinds of limestone to resolve why grass tetany (i.e., hypomagnesemic tetany) was a livestock problem in Michigan and not in Wisconsin. Except for the recognized differences in the underlying bedrock, which in many places also affected the glacial drift mantle above the bedrock, the soils in the two states are essentially alike in genesis and morphology. Samples of grasses from states adjoining Wisconsin and Michigan were included. The grasses grown on soils formed in or influenced by dolomitic limestone had 0.26% Mg, whereas similar species of grasses from soils influenced by calcitic limestone had 0.14% Mg. Plant Mg levels of 0.20% or more protect cattle against grass tetany.

The principal focus of the soil–plant studies has been on problems of critical deficiencies and toxicities affecting grazing animals. The concentration limits of a given trace element in feed and forage plants associated with acute deficiencies and toxicities are fairly well established by experimentation and by animal feeding trials and are useful guides. Areas associated with marginal, rather than acute, deficiencies and toxicities undoubtedly occur.

Marginally toxic areas of Mo for livestock can be expected to occur adjacent to the critical areas (Figure 6). In the case of Se (Figure 8), marginal deficiencies can reasonably be expected to occur in the variable-Se areas, and marginal or chronic toxicities near areas of acute Se toxicity.

In some parts of the western rangeland, marginal Zn deficiency may be a nutritional problem in grazing animals. Weight gains have been reported in range cattle fed supplemental Zn (Mayland *et al.*, 1980). Unpublished data (Kubota, 1983) suggest that Zn concentrations tend to be uniformly low in a wide range of plants when the plants are grown on western rangeland soils formed in or influenced by volcanic ash.

Often when the distribution of nutritional problem areas is investigated, interpretations about the role of soils and soil parent materials on the trace element concentration in plants are made on the basis of collective information from several different sources. For example, much of northern Missouri was included in an area of adequate Se (Figure 8). Although relatively few samples of plants were collected and analyzed, the Se concentrations were uniformly within the adequate Se range even though the samples were from an area of dominantly acid soils formed in loess and weathered glacial drift. The source of the soil Se was attributed to surficial materials transported and deposited by the Missouri River system. The Missouri River originates in the northern Plains States and drains areas where Cretaceous and Tertiary sedimentary rocks crop out on the landscape. The river was considered to be a major pathway when Se was redistributed southward in the United States by ice, wind, and rain. Similarly, earlier reports about Se-rich "nontoxic seleniferous" soils in Hawaii have been useful in evaluating reported cases of "yellow calf disease," a nutritional disorder attributed to excess Mo in feed and forage plants on the island of Hawaii (Campbell *et al.*, 1972). The soil Mo concentration in three samples of Oxisols from the island of Hawaii ranged from 13.5 to 18.6 $\mu g\ g^{-1}$ In comparison, Oxisols on the island of Oahu were found to have 1.8 to 2.2 $\mu g$ Mo per gram. The median Mo concentration in the United States soils was reported to be slightly more than 1 $\mu g\ g^{-1}$ and ranged from about 0.08 to over 30 $\mu g\ g^{-1}$ (Kubota, 1977).

Many of the Se concentrations in plants reported in the study of Se-toxic areas in the 1930s appears to have been determined in plants grown on well-drained soils, because these were the soils most often implicated with selenosis in the animals. The reported occurrence of seleniferous soils in a wet basin in Ireland (Fleming, 1962) suggests a need to determine possible levels and forms of organic soil Se that are likely to be associated with surface soil organic matter accumulation in the wet basin. The availability of Se to plants from wet soils could be partly temperature dependent and increase with rising soil temperature as well as with improved soil drainage. The high Se levels found in the wet Irish soils presumably came from Se that was initially present in the soils on the higher lying areas and subsequently leached out, redeposited, and concentrated in soils in the lower lying areas.

In general, more data are available on trace element content in plants than in soils on which the plants have grown. Such a disparity in the data base between plants and soils may reflect in part difficulties in determining soil trace element concentrations. Soils have many interfering elements that must be removed or corrected for before reliable measurements can be obtained of any given trace element in the soils. Selecting soil samples for laboratory study on the basis of plant concentrations is a means by which definitive samples of soils can be screened and selected for study. For example, it was observed that 40 to 50 μg Mo per gram were in the high concentration range of Mo in legumes from soils formed in alluvium of shale origin. By comparison, concentrations of 100 to 400 μg Mo per gram were in the high range in similar species of legumes grown on soils formed in granitic alluvium. When Mo was determined in selected samples from the two groups of soils, the soils of shale rock origin were found to have more Mo than the soils of granite rock origin (Kubota, 1977). In granite, the Mo is reported to be present as molybdenite but a specific mineral carrier of Mo in shale has not been identified.

Similar differences have been observed when soil Co (total) was compared in sandy soils of the Southeastern and Northeastern United States. Both groups of sandy soils have been associated with low Co (0.04 to 0.07 $\mu g\ g^{-1}$ or less) forage plants. The sandy Humaquods formed in Coastal Plain deposits of the Southeast had 1 $\mu g\ g^{-1}$ or less of total Co, compared with 2 to 3 μg Co per gram in Spodosols of New England, which were formed in granitic glacial drift (Kubota and Lazar, 1960). The differences in Co concentrations between the sandy soils were also evident among soils of heavier textures. Such observations suggest that much of the Co in the soils of New England still remains in primary minerals and in forms unavailable for uptake by plants. Studies of extractable forms of soil Co supports this view (Kubota, 1965). While the soils of the Southeastern United States included in the study had less total Co, the amounts in forms extractable with dithionite and acetic acid–dithizone were greater than from the soils formed in glacial drift of the Northeast. The distribution pattern of total, dithionite-extractable, and acetic acid–dithizone-extractable Co with soil depth in Ultisols suggested that the Co in these soils might be recycled biologically and thereby maintain the available Co status of the soils.

Generally, the water of natural streams and rivers, although a potential source, appears to be a poor source of soil trace elements. Continued use and reuse of water downstream for irrigation, however, may result in an accumulation of a trace element in soils downstream. Increases in Se levels in downstream waters of the Colorado River and its tributaries has been attributed to the use and reuse of the water for irrigation in an arid region (Lakin, 1961b).

Unlike water of streams and rivers, however, the waters of some thermal springs in Nevada are significant sources of some trace elements that can accumulate in the soil. High concentrations of $F^-$ (Kubota *et al.*, 1982) and of B (Garside and Shilling, 1979; Kubota, unpublished) have been determined in water of some thermal springs in Nevada. A buildup of salts of B and $F^-$ in soils is quite likely, because little water flowing into or originating in Nevada flows out of the state through major drainage systems.

Rain, mist, and fog carried inland by wind from oceans may be an important source of dietary I for animals and man, as suggested by a low incidence of goiter near coastal areas. Decreases of 10- to 100-fold in I levels in rain water have been reported with increasing distance inland from coastal areas to interior parts of New England and New York (McClendon, 1939). Selenium in rain water has been determined in samples collected from eight locations across the United States from west to east (Kubota *et al.*, 1975). An effect of the rain water on the Se concentration of common forage plants grown in New York was not evident. The small amount of Se deposited by rainfall on any one of three annual cuttings of a forage crop may be a factor. Of probable greater importance was the amount of Se present in insoluble particulates. About one half of the Se was in such particulates. The Se in particulates may be largely unavailable for plant uptake.

Regional patterns of Se and Mo plant concentrations have been used to evaluate the magnitude of trace element movement in the food and feed chain of man. For example, effects of Se fertilization on the Se concentration in alfalfa (*Medicago sativa* L.) was determined by growing alfalfa with and without Se applied to a soil in a low-Se area. The hay was fed to ewes and the transfer of the Se was measured in the lambs (Allaway *et al.*, 1966). Lambs from ewes fed the Se-fertilized hay had a lower incidence of white muscle disease than did lambs from ewes fed hay from the unfertilized field. The transfer of the applied Se from soil to plant was also reflected in Se concentrations in liver, kidney, and muscle of the lambs from ewes fed the Se-fertilized hay. More Se was found in processed and unprocessed milk from high-Se areas than in processed and unprocessed milk from low-Se areas in the United States (Allaway *et al.*, 1968). Whole blood samples of residents living in low- and adequate-Se areas, however, did not show a consistent pattern in Se levels. The samples of human blood were not from the same locations as those of the samples of milk.

Information about the geographical distribution of Mo-toxic areas for grazing animals has served as a basis to assess dental decay rates of children born and raised in high-Mo areas in New Zealand (Ludwig *et al.*, 1960), the United Kingdom (Anderson, 1966), and Hungary (Adler and Staub, 1953). The results of these studies indicated that children born and raised in high-Mo areas have good teeth, based on numbers of decayed, missing, and filled teeth (DMFT). A similar study was conducted in the United States. The DMFT rates in children were determined in a series of towns paired on the basis of population size and comparable levels of

$F^-$ in the domestic water supply by a single research dentist. This minimized variability in observations (Curzon *et al.*, 1971). When studied in this manner, no significant differences in DMFT rates were evident in children from the series of paired towns. The DMFT rates, however, were low, suggesting that a generally high background level of Mo may have masked out differences in DMFT rates in children of the paired towns. Like Se, samples of processed and unprocessed milk from selected high-Mo areas had more Mo than samples from outside of the high-Mo areas (Allaway *et al.*, 1968). The set of blood samples, on which Se was determined earlier, did not differ in Mo.

## V. Future Outlook

The many trace elements currently recognized to be nutritionally important to animals and man suggest a wide range of beneficial soil–plant studies that could be undertaken. In many cases, findings from such research contribute to an understanding about the environmental behavior of the trace element as well. The problem generally seems to be one of judicious selection of the meaningful studies rather than a lack of them. The increasing availability, for agricultural research, of multielement analytical instruments with capabilities to simultaneously determine microgram and nanogram amounts of a number of trace elements in biological materials provides many opportunities for new studies. An extremely large amount of useful information resulted from the application of atomic absorption analyses to studies of trace elements in soils and plants, and similar surges may result as plasma emission spectrometry becomes increasingly available to soil and plant scientists. Its application could help resolve many questions about the impact of mineral interactions on uptake and utilization of mineral elements by plants. A number of two-way interactions between nutritionally important elements have been recognized. Examples include Al with Ca, Mg, Fe, and Zn; K with Ca and Mg; Ca–Mg; and Zn with Cd, Cu, and P. In addition to two-way interactions, three- and four-way mineral–mineral interactions also occur, e.g., Al–Fe–Zn–P. Multielement analyses offer a powerful tool for identifying such interactions and for assessing their impact on the uptake and distribution of mineral elements in plants and animals.

The versatility of inductively coupled plasma emission spectrometry (Fassell, 1978) was demonstrated, for example, in a study of how Cd, Pb, Zn, and Cu were distributed in reproductive parts of selected seed and grain crops (Kubota and Welch, manuscript in preparation). The following parts were separated, ground, and analyzed separately: corn—husks, silk and grain; wheat, oats, and barley—grain, rachilla, and glumes; and soybean—seed, pods, and stems. When simultaneously determined, the concentration of Zn was highest and that of Cd and Pb were the lowest in the seed and grain compared with the other parts of the plant. The role of the plant in the distribution of these four trace elements among various plant parts greatly exceeded any role that could be attributed to soils on which the plants were grown.

The availability of advanced instrumentation for research, however, points toward a need for greater care in sample selection and collection in soil–plant studies. The magnitude of sampling error in field studies, in particular, far exceeds analytical errors, which can be controlled in most laboratories. Identifying the soils on which the plants are grown seems essential, because the information about soils provides a means by which geographical distribution patterns in plants can eventually be depicted in map form.

Information about nutritionally important trace elements is broadly based and includes reported observations from many disciplines, such as animal and veterinary sciences, biochemistry, animal and human nutrition, as well as from soil and plant sciences. Such background information provides a valuable frame of reference from which soil and plant scientists can select relevant studies that may lead to an improved understanding of the behavior of trace elements in the feed and food chain of animals and man. A study of Cr in plants illustrates one possible multidisciplinary study. Evidence exists to show clearly that diabetics and laboratory animals respond to Cr supplements. Therefore, Cr in food and feed plants may be an important source of Cr for man and animals. However, the biologically active form of Cr is recognized as the glucose tolerance factor (Mertz, 1981), and establishment of its relationship to plant Cr requires research on both the amounts and the naturally occurring forms of Cr in plants.

Increased knowledge about soil–plant relationships of trace elements could also be combined with considerations about the bioavailability of mineral elements in plants (Welch and House, 1984) and presence of naturally occurring "antinutrients" in edible crops (House and Welch, 1984). This information could provide a better understanding of how different elements behave in the complex food and feed chain of man and animals as well as in the geochemical environment (Kubota, 1983b).

## References


Adler, P., and J. Staub. 1953. A waterborne caries protective agent other than fluorine. *Acta Med. Acad. Sci. Hung.* 4:221.

Allaway, W.H., J. Kubota, F. Losee, and M. Roth. 1968. Selenium molybdenum and vanadium in human blood. *Arch. Environ. Health* 16:342–348.

Allaway, W.H., D.P. Moore, J.E. Oldfield, and O.H. Muth. 1966. Movement of physiological levels of selenium from soils through plants to animals. *J. Nutrit.* 88:411–418.

Anderson, R.J. 1966. Dental caries prevalence in relation to trace elements. *Br. Dent. J.* 120:271–275.

Anderson, A.J., and E.J. Underwood. 1959. Trace element deserts. *Sci. Am.* 200:97–106.

Andrews, E.D. 1956. Cobalt deficiency. New Zealand *J. Agron.* 92:239–244.

Beeson, K.C. 1961. Occurrence and significance of selenium in plants. In: Selenium in Agriculture. USDA Agric. Handbook 200. US Government Printing Office, Washington, DC, pp. 34–40.

Berger, K.C. 1962. Micronutrient deficiencies in the United States. *J. Agric. Food Chem.* 10:178–181.

Campbell, C.M., J.C. Nolan, Jr., R.W. Stanley, R.M. Nakamura, and O. Wayman. 1972. "Yellow calf" can be corrected. *Hawaii Farm Sci.* 21:4–6.

Carlisle, E.M. 1974. Essentiality and function of silicon. In: W.G. Hoekstra, J.W. Suttie, H.E. Ganther, and W. Mertz (eds.) *Trace Element Metabolism in Animals—2.* University Park Press, Baltimore, MD, pp. 407–424.

Curzon, M.D.J., J. Kubota, and B.G., Bibby. 1971. Environmental effects of molybdenum on dental caries. *J. Dent. Res.* 50:74–77.

Eskew, D., R.M. Welch, and E. Cary. 1983. Nickel: an essential micronutrient for legumes and possibly all higher plants. *Science* (Washington, DC) 222:621–623.

Fassel, V.A. 1978. Quantitative elemental analyses by plasma emission spectroscopy. *Science* (Washington, DC) 202:183–191.

Fleming, G.A. 1962. Selenium in Irish soils and plants. *Soil Sci.* 94:28–35.

Ferguson, W.S., A.H. Lewis, and S.J. Watson. 1943. The teart pastures of Somerset. *J. Agric. Sci.* 33:444–451.

Garside, L.J., and J.H. Schilling. 1979. Thermal waters of Nevada. Nevada Bureau of Mines and Geology, Bull 91. University of Nevada, Reno, NV. p. 164.

Goldschmidt, V.M. 1954. *Geochemistry.* The Clarendon Press, Oxford, p. 730.

House, W.A., and R.M. Welch. 1984. Effects of naturally occurring antinutrients on the nutritive value of cereal grains, potato tubers and legume seeds. In: R.M. Welch and W.H. Gabelman (eds.) *Crops as Sources of Nutrients for Humans.* American Society of Agronomy, Crop Science Society of America, and Soil Science Society of America, Spec. Publ. 48. American Society of Agronomy, Madison, WI, pp. 9–35.

Howarth, R.J., and I. Thornton. 1983. Regional geochemical mapping and its application to environmental studies. In: I. Thornton (ed.) *Applied Environmental Geochemistry.* Academic Press, New York, pp. 41–70.

Kubota, J. 1964. Cobalt content of New England soils in relation to cobalt levels in forages for ruminant animals. *Soil Sci. Am. Proc.* 28:246–251.

Kubota, J. 1965. Distribution of total and extractable forms of cobalt in morphologically different soils of Eastern United States. *Soil Sci.* 99:166–174.

Kubota, J. 1968. Distribution of cobalt deficiency in grazing animals in relation to soils and forage plants of the United States. *Soil Sci.* 106:122–130.

Kubota, J. 1972. Sampling of soils for trace element studies. *Ann. N.Y. Acad. Sci.* 199:105–177.

Kubota, J. 1975. Areas of molybdenum toxicity to grazing animals in the Western States. *J. Range Mgmt.* 28:252–256.

Kubota, J. 1977. Molybdenum status of U.S. soils and plants. In: W. Chappell and K. Peterson (eds.) *The Geochemistry, Cycling and Industrial Uses of Molybdenum.* Vol. II. Marcel Dekker, New York, pp. 555–581.

Kubota, J. 1983a. Copper status of United States soils and plants. *Agron. J.* 75:912–918.

Kubota, J. 1983b. Soils and plants and the geochemical environment. In: I. Thornton (ed.) *Applied Environmental Geochemistry*. Academic Press, New York, pp. 103–122.

Kubota, J., and W.H. Allaway. 1972. Geographic distribution of trace element problems in the United States. In: J.J. Mortvedt, P.M. Giordano, and W.L. Allaway. 1972. Geographic distribution of trace element problems in the United States. In: J.J. Mortvedt, P.M. Giordano, and W.L. Lindsay (eds.) *Micronutrients in Agriculture*. Soil Science Society of America, Madison, WI, pp. 525–554.

Kubota, J., and V.A. Lazar. 1958. Cobalt status of soils of Southeastern United States. II. Groundwater Podzols and six geographically associated soil groups. *Soil Sci.* 86:262–268.

Kubota, J., and V.A. Lazar. 1960. Cobalt-iron relationship in sandy soils. *Trans. 7th Intl. Congr. Soil Sci.* 2:134–140.

Kubota, J., W.H. Allaway, D.F. Carter, E.E. Cary, and V.A. Lazar. 1967a. Selenium in crops in the United States in relation to selenium-responsive diseases in animals. *J. Agric. Food Chem.* 15:448–453.

Kubota, J., E.E. Cary, and G. Gissel-Nielsen. 1975. Selenium in rainwater of the United States and Denmark. In: D.D. Hemphill (ed.) *Trace Substances in Environmental Health—IX*. University of Missouri Press, Columbia, MO, pp. 123–130.

Kubota, J., V.A. Lazar, and K.C. Beeson. 1960. The study of cobalt status of soils in Arkansas and Louisiana using the black gum as the indicator plant. *Soil Sci. Soc. Am. Proc.* 24:527–528.

Kubota, J., V.A. Lazar, L.N. Langan, and K.C. Beeson. 1961. The relationship of soils to molybdenum toxicity in cattle in Nevada. *Soil Sci. Soc. Am. Proc.* 25:227–232.

Kubota, J., V.A. Lazar, G.H. Simonson, and W.W. Hill. 1967b. The relationship of soils to molybdenum toxicity in grazing animals in Oregon. *Soil Sci. Soc. Am. Proc.* 31:667–671.

Kubota, J., E.R. Lemon, and W.H. Allaway. 1963. The effect of soil moisture upon the uptake of molybdenum, copper and cobalt by alsike clover. *Soil Sci. Soc. Am. Proc.* 27:679–683.

Kubota, J., E.A. Naphan, and G.H. Oberly. 1982. Fluoride in thermal spring water and in plants of Nevada and its relationship to fluorosis in animals. *J. Range Mgmt.* 35:188–192.

Kubota, J., G.H. Oberly, and E.A. Naphan. 1980. Magnesium in grasses of three selected regions in the United States and its relation to grass tetany. *Agron. J.* 72:907–914.

Lakin, H.W. 1961a. Vertical and lateral distribution of selenium in sedimentary rocks of Western United States. In: Selenium in Agriculture. USDA Agric. Handbook 200. US Government Printing Office, Washington, DC, pp. 12–24.

Lakin, H.W. 1961b. Geochemistry of selenium in relation to agriculture. In: Selenium in Agriculture. USDA Agric. Handbook 200. US Government Printing Office, Washington, DC, pp. 3–12.

Ludwig, T.G., W.B. Healy, and F.L. Losee. 1960. An association between dental caries and certain soil conditions in New Zealand. *Nature* (London) 186:695.

Mayland, H.F., R.C. Rosenaw, and A.R. Florence. 1980. Grazing cow and calf responses to zinc supplementation. *J. Anim. Sci.* 51:966–974.

McClendon, J.F. 1939. *Iodine and the Incidence of Goiter*. University of Minnesota Press, Minnespolis, MN.

Mertz, W. 1969. Chromium occurrence and function in biological systems. *Physiol. Rev.* 49:163–239.

Mertz, W. 1981. The essential trace elements. *Science* (Washington, DC) 213:1332–1338.

Muth, G.H., and W.H. Allaway. 1963. The relationship of white muscle disease to the distribution of naturally occurring selenium. *J. Am. Vet. Med. Assoc.* 142:1379–1384.

National Academy of Sciences (Committee on Biologic Effects of Atmospheric Pollutants). 1971. Fluorides. National Academy of Sciences, Washington, DC, p. 285.

Nielsen, F.H. 1974. Essentiality and function of nickel. In: W.H. Hoekstra, J.W. Suttie, H.E. Ganther, and W. Mertz (eds.) *Trace Element Metabolism in Animals—2*. University Park Press, Baltimore, MD, pp. 381–396.

Nielsen, F.H. 1984. Ultratrace elements in nutrition. *Ann. Rev. Nutrit. 1984* 4:21–41.

Olson, O.E., E.I. Whitehead, and A.L. Moxon. 1942. Occurrence of soluble selenium in soils and its availability to plants. *Soil Sci.* 54:47–53.

Rankama, K., and T.G. Sahama. 1950. *Geochemistry*. University of Chicago Press, Chicago, p. 912.

Robinson, W.O., and G. Edgington. 1954. Availability of soil molybdenum as shown by the molybdenum content of many different plants. *Soil Sci.* 77:237–251.

Robinson, W.O., G. Edgington, W.H. Armiger, and A.V. Breen. 1951. Availability of molybdenum as influenced by liming. *Soil Sci.* 72:267–274.

Shacklette, H.H., J.C. Hamilton, J.G. Boerngen, and J.M. Bowles. 1971. Element composition of surficial materials in the conterminous United States. US Geological Survey, Prof. Paper 574-D, pp. 1–71.

Soil Conservation Service. 1975. Soil taxonomy. A basic system of soil classification for making and interpreting soil surveys. USDA Agric. Handbook 436. US Government Printing Office, Washington, DC, p. 754.

Staker, E.V. 1943. Progress report on the control of zinc toxicity in peat soils. *Soil Sci. Soc. Am. Proc.* 7:387–392.

Thacker, E.J. 1961. Effect of selenium on animals. In: Selenium in agriculture. USDA Agric. Handbook 200. US Government Printing Office, Washington, DC, pp. 46–53.

Thornton, I. 1983. Geochemistry applied to agriculture. In: I. Thornton (ed.) *Applied Environmental Geochemistry*. Academic Press, New York, pp. 231–263.

Underwood, E.J. 1971. *Trace Elements in Human and Animal Nutrition*. 3rd ed. Academic Press, New York, 543 p.

Welch, R.M. 1985. Nutrition. In: *McGraw-Hill 1986 Yearbook of Science and Technology*. McGraw-Hill Book Co., New York, pp. 320–322.

Welch, R.M., and W.A. House. 1984. Factors affecting the bioavailability of mineral nutrients in plant foods. In: R.M. Welch and W.H. Gabelman (eds.) *Crops as Sources of Nutrients for Humans*. American Society of Agronomy, Crop Science Society of America, and Soil Science Society of America. Spec. Publ. 48. American Society of Agronomy, Madison, WI, pp. 37–54.

# Index


**A**

Actinium, 101, 102
Adularia, 10
Aegirine, 115
Alabama, 180
Albany ail, 193
Albite, 109, 111, 113, 115
Alfalfa, 24, 42, 150, 153, 167, 181, 185, 210
Alfisols, 10, 14, 32, 78, 119
Alkali disease, 198
Alkali elements, 101–138
Allophane, 9, 13
Alluvial soils, 120, 122
Aluminum, 4, 10, 11, 13, 15, 23, 35, 37, 122, 211
Alunite, 5, 15, 16
Amazonite, 116
Amblygonite, 114
Ammonium, 3, 5, 7, 33, 128–130
Amphiboles, 109, 114, 115, 118, 136
Analcime, 115
*Ananas comosus* L. Merr., 185
Andosols, 10
*Andropogon virginicus* L., 196
Anorthite, 109, 115
Aquic Hapludoll, 75, 80
Arenic Hapludult, 39
Arenic Paleudult, 75
Aridic Haploborol, 12
Aridic soils, 119
Aridosols, 119
Arrhenius equation, 27, 28, 48
Artificial isotopes, 101, 123
*Astragalus bisulcatus* Hook., 204
Australia, 135, 190, 193
*Avena sativa*, 185
Avogadrite, 117
Azonal soils, 121

**B**

Ballontini, 81
Barium, 4, 5, 11, 29, 116
Barley, 80, 81, 128, 211
Basalt, 4, 111
Beidellite, 14
Beryl, 117
Beryllium, 11
Biological activity, 7, 34, 43
Biotite, 7, 9, 10, 12, 14, 24, 34, 35, 37–39, 43, 109, 112, 115–117, 128, 129, 137, 138
Black gum, 195–197
Borates, 114
Boron, 150, 167, 191, 210
Boussinesq's theory, 90
British soils, 23, 120, 122
Bronsted's reaction rate theory, 28
Brown Forest soils, 120–122
Bulk density, 26, 71, 72, 75, 76, 77, 79, 80, 82, 83, 91–96

**C**

Cadmium, 191, 194, 202, 211
Calcium, 4, 5, 7, 11, 13–17, 20, 29, 33, 44, 46–48, 111, 115, 127, 150,

Calcium (*continued*)
152, 153, 164, 167, 169, 170, 171, 174, 175, 177–179, 181, 182, 211
California, 199, 201
*Camelia senensis* L. Ktze, 185
Canada, 180, 203
Cation exchange capacity, 3, 9, 13–15, 41, 48, 182
Cation exchange sites, 106
Cattle, 192, 197
Cecil soil, 119
Cesium, 4, 5, 11, 29, 101, 102, 105, 107, 108, 110–112, 116–121, 123, 125, 126, 128, 130, 137
Chernozem soils, 29, 119–122
Chester soil, 17, 18, 50, 51
Chestnut soils, 119
Chlorite, 6, 12, 14
Chocorua's curse, 193
Chromium, 193
Citrus, 103, 185
*Citrus sinensis* L., 178, 185
Cobalt, 191, 193–199, 202, 206, 209
Colorado, 121, 134
Compactibility, 68
Compaction, soil, 65–98
Compressibility, 67, 68
Compression, 66, 67, 75, 77
Copper, 11, 171, 191, 192, 194, 196, 201–203, 206, 211
Corn, *also see* Maize, 8, 150, 151, 153, 155, 158, 163, 165, 168, 169, 171, 172, 174, 175, 177, 179, 183, 185, 192, 211
Cotton, 80
Chromium, 191, 194, 202, 212
Crop residues, 122, 129, 133
Cryolithionite, 114

**D**

Delaware, 7, 8, 12, 30, 38, 39, 41, 44, 180
Diffusion, 7, 18, 19, 23, 37, 47, 49, 50, 51, 78, 132
Diopside, 115
Dispersion coefficient, 26
Doland loam, 93
Dothan soil, 30
Downer sandy loam, 49–51
Durham soil, 119

**E**

Earthworms, 121
Edina soil, 134
Electronegativity, 105, 106
Electroultrafiltration, 6, 31, 32
Elovich equation, 19, 22
Entisols, 10, 77, 119
Equivalent conductivity, 127
Erosion, 109, 123
Evaporation, 121
Evapotranspiration, 84, 85
Evesboro soil, 20, 21
Exchangeable sodium percentage, 103, 135
Eyring's reaction rate theory, 44, 45

**F**

Feldspar, 5, 7–12, 36, 37, 103, 109, 111–113, 115–118, 120, 126, 131, 136
Feldspathoids, 115
First-order equation, 22
First-order plot, 21
First-order reaction, 20, 26, 31
Florida, 30, 179
Fluorine, 191, 206, 209, 210, 211
Food chain, 103, 212
Francium, 101, 102
Frohlich's equations, 90–93

**G**

Gabbro, 111
Gallium, 11
Gaussian distribution, 155
Georgia, 180
German soils, 120
Ghanaian forest soils, 122
Gibbs-Duhem equation, 46
Gibbsite, 22
Glauker, J.R., 4
Gley soils, 121, 122
*Glycine max* L. Merrill, 178, 185, 192
Goethite, 22
*Gossypium hirsutum* L., 80
Grand Traverse disease, 193
Granite, 4, 10, 111, 122, 195, 209
Grass tetany, 206, 207
Grey Forest soils, 119
Ground water, 103, 110

**H**
Halloysite, 15
Haverhill soil, 12
Hawaii, 180, 208
Hectorite, 114
*Helianthus annus* L., 185
*Hevia brasiliensis* H.B.K., 185
Histosols, 119
Holland, 3, *also see* The Netherlands
Hooke's law, 88
*Hordeum vulgare* L., 80
Hornblende, 14
Humaquods, 196, 209
Hungary, 210
Hydraulic conductivity, 97

**I**
Idaho, 199
Igneous rocks, 4, 10, 11, 109–115, 119, 121
Illinois, 174, 180, 205
Illite, 6, 10, 43, 114–117, 124, 128, 130
Inceptisols, 10, 77, 119
Indiana, 28, 180
Infiltration rate, 82–84
Interzonal soils, 121
Iodine, 191, 193
Ionic potentials, 106
Ionization potential, 105, 106
Iowa, 133, 180
Iron, 35, 111, 113, 114, 171, 191, 192, 211
Irrigated conditions, 121
Irrigation, 192, 209
Irrigation water, 103, 126, 127
Israel, 93, 98

**J**
Japanese soils, 13
Jarosite, 137

**K**
Kalmia sandy loam, 39, 40
Kaolinite, 9, 15, 17, 18, 20, 21, 29, 30, 114–117
Karahoa pumice, 197
Kennansville soil, 39, 40
Kentucky, 205

**L**
Lambs, 210
Leaching, 3–5, 7, 13, 26, 28, 118, 126, 131, 196
Lead, 11, 191, 194, 211
Lepidolite, 114–117, 129
Lithium, 4, 5, 11, 101–103, 105–108, 110–115, 118–123, 125–127, 130, 135
Lime, 174
Limestone, 4, 10, 110–112, 200, 206, 207
Lloyd clay, 73

**M**
Magnesium, 4, 5, 7, 14, 16, 33, 35, 111, 113, 114, 127, 150, 152, 153, 164, 167, 169–171, 174, 175, 177–179, 181, 182, 206, 207, 211
Maine, 196, 198
Maize, *also see* Corn, 156, 180
Malic Haploxeralf, 93
Manganese, 35, 150, 171, 191–193
Manure, 123
Marshall soil, 131–134
Maryland, 180
Massachusetts, 193
Matapeake soil, 20, 30, 45
*Medicago sativa* L., 24, 42, 150, 185, 210
Metamorphic rocks, 10, 11
Mica, 3, 5, 6, 8, 10, 12, 13, 15, 29, 32–34, 42, 43, 103, 111, 113, 115, 116, 118, 120, 128, 129, 138
Michigan, 180, 193, 206, 207
Microbial activity, 79
Microcline, 7, 9–11, 38, 39, 43, 115, 116
Minnesota, 93, 180, 205
Miscible displacement, 17–20, 44, 49, 51
Mississippi, 204
Missouri, 121, 204
Molybdenum, 191, 192, 194–196, 199–203, 206, 208–211
Modeling, 3, 24, 52
Models, soil compaction, 65–98
Mollisols, 10, 77, 119
Montmorillonite, 3, 14, 20, 21, 29, 30, 107, 114–117

Muck, 195
Muscovite, 7, 9, 10, 12, 13, 34, 35, 37–39, 43, 44, 109, 113–117, 128, 129, 135

N
Nahal Oz silt loam, 93
Natural isotopes, 101
Neck's ail, 193
Nephaline, 115
Nevada, 191, 198, 199, 204, 209, 210
New Guinea, 135
New Hampshire, 193, 196
New Jersey, 180
New York, 210
New Zealand, 121, 180, 193, 197, 210
Nickel, 193
Nitrate, 150
Nitrifying bacteria, 128
Nitrogen, 24, 118, 122, 150, 152–160, 163–179, 181–184
Nitrogen fixation, 193
Nontorite, 14
North Carolina, 121, 180, 196
*Nyssa sylvatica biflora* (Walt) Sarge., 196

O
Oats, 185, 211
Ohio, 180
Olivine, 109, 113, 114
Oranges, 178
Oregon, 197, 199, 200, 204
Organic matter, 5–7, 16, 20, 75, 122, 123, 136, 198
Organic carbon, 74
Orthoclase, 6, 7, 9–11, 38, 39, 44, 113, 115
*Oryza sativa* L., 193
Oxisols, 10, 32, 78, 119, 208
Oxygen diffusion, 97

P
pH, 43, 130, 201
Parabolic diffusion equation, 19, 22, 23
Paragonite, 115, 135
Pea, 80
Peaches, 150, 151, 185
Peat, 123, 195
Pegmatite, 111, 117
Penetrometer, 80, 81
Pennsylvania, 180
Perisdotites, 4
Petalite, 114
Phlogopite, 10, 34, 35, 37
Phosphate, 114
Phosphorus, 22, 118, 150, 152–160, 163–180, 181–184, 211
Phyllite, 122
Pigs, 192
Pineapple, 171, 185
*Pisum sativum* L., 80
Plagioclase, 111, 113–116
Plinthic Paleudults, 30
Podzols, 119, 121, 122
Poisson coefficient, 90
Poisson ratio, 88
Pollucite, 111, 116, 117
Polylithionite, 114
Poplars, 185
*Populus spp.*, 185
*Pore water velocity*, 26, 27
Potassium, 1–82, 101–108, 110–113, 115–125, 127–132, 136–138, 150, 152, 153, 156, 158, 159, 163–179, 181–184, 211
Potassium nitrate, 4
Potato, 150, 171, 172, 173, 185
Power-function equation, 19, 23, 24
*Prunus persica* Batsch, 150, 185
Pyroxene, 109, 113–115, 118, 136

Q
Quartz 10, 74, 113, 120

R
Radioactive fallout, 104
Radioactive isotopes, 101, 104
Rainfall, 110
Rainfall intensity, 83, 84
Rubidium, 4, 5, 11, 101, 102, 105, 107, 108, 110–112, 116–123, 125, 126, 128, 130, 137
Red soils, 119, 120, 122
Rhizosphere, 43

Rhodesia (Zimbabwe), 120
Rhodizite, 117
Rhyolite, 11, 111
Rice, 193
Rodents, 121
Royal Agricultural Society, 28
Rubber, 185
Runoff, 110
Russia, 3, 120, 122
Ryegrass, 32

S
*Saccarum officinarum* L., 150, 185, 193
Sandstone, 110–112, 122
Sanidine, 9–11, 36
Scottish soils, 121, 123
Sedimentary rocks, 4, 10, 11, 110, 113, 115, 204, 208
Selenium, 191, 193, 194, 195, 202, 205, 207–211
Serozem soils, 119
Serpentinitic rocks, 10
Sewage sludge, 123
Shale, 110, 112, 124, 195, 200, 201, 204, 209
Sheep, 192
Sialic rocks, 10
Siderophyllite, 114
Sierozem soils, 127
Silicon, 3, 4, 10, 11, 13, 34, 108, 112, 113, 122, 183, 191, 193
Smectite, 9, 14, 15, 114
Sodalite, 115
Sodium, 4, 5, 11, 13, 14, 43, 101–103, 105, 107, 108, 110–116, 118, 121–123, 125, 127, 128, 130, 131, 133, 135, 137, 183
Sodium absorption ratio, 127, 135
Soil aggregates, 78, 79, 83
Soil development, 106
Soil structure, 74
*Solanum tuberosum* L., 150, 185
Solonetz soils, 121
South Africa, 179, 180, 182
South Carolina, 196
Soybean, 178, 181, 185, 192, 193, 211
Spodosols, 77, 119, 195, 196, 209
Spodumene, 114
*Stanleya bipinnata* (Pursh.) Britt, 204
Strontium, 4
Sugar beet, 103
Sugarcane, 150, 165, 170–172, 176, 182, 185, 193
Sulfate, 15, 16
Sulfur, 150, 166–170
Sunflower, 185
Switzerland, 24
Syenite, 4

T
Taranakite, 5, 15
Taupo ash, 197
Tea, 185
Tennessee, 205
Termites, 121
Texas, 196
Thalium, 11, 116
The Netherlands, 4, *also see* Holland
Titanium, 113
Tillage, 65, 133
Tin, 191, 194
Trachyte rocks, 11
*Triticum aestivum* L., 175, 185
Typic Haplaquoll, 83, 95, 96
Typic Hapludoll, 76, 93
Typic Hapludult, 17, 39, 73
Typic Haplustox, 75
Typic Hydrandept, 75
Typic Quartzipsamment, 20
Typic Xerothent, 85

U
Udic Haploboroll, 93
Udic soils, 119
Ultisols, 10, 14, 78, 119, 209
United Kingdom, 210
United States, 3, 66, 98, 113, 118–120, 122–124, 133, 179–181, 189, 190, 192–196, 201–210
Ustic soils, 119

V
van't Hoff equation, 48
Vanselow selectivity coefficient, 46

Vermiculite, 3, 5–7, 9, 12–15, 17, 18, 20, 21, 28, 30, 33, 44, 115
Vertisols, 77
Virginia, 30, 180, 200
Volcanic ash, 204
von Liebig, 4

**W**
Washington, 199
Waukegan silt loam, 76, 93, 94
Webster clay loam, 83, 95, 96
Wheat, 175, 183, 185, 211
Wiley soil, 134
Wisconsin, 180, 206, 207
Wollast model, 36
Wyoming, 121

**X**
Xeric soils, 119

**Y**
Yellow calf disease, 208
Yellow soils, 120, 122

**Z**
*Zea mays* L., 8, 150, 185, 192
Zeolite, 5, 13, 14, 115, 137
Zero-order equation, 19, 22
Zinc, 150, 171, 191, 192, 208, 211
Zinnwaldite, 114, 116
Zonal soils, 121

MIX
Papier aus verantwortungsvollen Quellen
Paper from responsible sources
FSC® C105338

If you have any concerns about our products,
you can contact us on
**ProductSafety@springernature.com**

In case Publisher is established outside the EU,
the EU authorized representative is:
**Springer Nature Customer Service Center GmbH**
**Europaplatz 3, 69115 Heidelberg, Germany**

Printed by Libri Plureos GmbH
in Hamburg, Germany